SUSTAINABLE DEVELOPMENT

Volume 19

Women and the Environment in the Third World
Alliance for the future

T0359067

Full list of titles in the set
SUSTAINABLE DEVELOPMENT

Volume 1: Global Environment Outlook 2000
Volume 2: Mountain World in Danger
Volume 3: Vanishing Borders
Volume 4: Atlas of Nepal in the Modern World
Volume 5: Caring for the Earth
Volume 6: Community and Sustainable Development
Volume 7: One World for One Earth
Volume 8: Strategies for National Sustainable Development
Volume 9: Strategies for Sustainability: Africa
Volume 10: Strategies for Sustainability: Asia
Volume 11: Strategies for Sustainability: Latin America
Volume 12: Wasted
Volume 13: Dam the Rivers, Damn the People
Volume 14: Rising Seas
Volume 15: Tomorrow's World
Volume 16: Social Change and Conservation
Volume 17: Threats Without Enemies
Volume 18: Progress for a Small Planet
Volume 19: Women and the Environment in the Third World
Volume 20: World at the Crossroads

Women and the Environment in the Third World
Alliance for the future

Irene Dankelman and Joan Davidson

publishing for a sustainable future

London • New York

First published in 1989

This edition first published in 2009 by Earthscan

ISBN 978-1-84407-950-6 (hbk Volume 19)
ISBN 978-0-415-85215-9 (pbk Volume 19)
ISBN 978-1-84407-931-5 (Sustainable Development set)
ISBN 978-1-84407-930-8 (Earthscan Library Collection)

First issued in paperback 2013

For a full list of publications please contact:

Earthscan
2 Park Square, Milton Park, Abingdon, Oxon OX14 4RN
Simultaneously published in the USA and Canada by Earthscan
711 Third Avenue, New York, NY 10017
Earthscan is an imprint of the Taylor & Francis Group, an informa business

Earthscan publishes in association with the International Institute for Environment and Development

A catalogue record for this book is available from the British Library

Library of Congress Cataloging-in-Publication Data has been applied for

Publisher's note
The publisher has made every effort to ensure the quality of this reprint, but points out that some imperfections in the original copies may be apparent.

WOMEN AND ENVIRONMENT IN THE THIRD WORLD

Alliance for the Future

Irene Dankelman and Joan Davidson

Earthscan Publications Ltd, London
in association with IUCN

First published 1988 by
Earthscan Publications Limited,
120 Pentonville Road, London N1 9JN,
E.mail: earthinfo@earthscan.co.uk
Website: http://www.earthscan.co.uk
in association with
The International Union of Conservation of Nature
and Natural Resources (IUCN)

Second printing 1989
Reprinted 1993, 1994, 1997

British Library Cataloguing in Publication Data

Women and environment in the Third
    World: alliance for the future.
    1. Developing countries. Environment.
    Conservation. Role of women
    I. Dankelman, Irene   II. Davidson, Joan
    333.7'2'091724
ISBN 1-85383-003-8

Typeset in Times Roman by
DP Phototypesetting, Aylesbury, Bucks

Front cover photograph: Maggie Murray/Format
Back cover photograph: Donald McCullin
Cover Design: David King

CONTENTS

Dedication   *v*
Authors' Preface   *vii*
Foreword by The Hon. Mrs Victoria Chitepo   *ix*
Introduction: Women have something to say   *xi*

**PART I   WOMEN, ENVIRONMENT AND NATURAL
         RESOURCES**

1. WHY WOMEN?   3

2. LAND: WOMEN AT THE CENTRE OF THE FOOD
    CRISIS   7
    Case Studies: Women and soybeans in Togo, West Africa   22
                   The Pinabetal Women's Organization,
                      Mexico   24
                   The Vacaria Project, Brazil   24
                   Tribal women in Iran   26
                   Golgotta Settlement, Ethiopia   27

3. THE INVISIBLE WATER MANAGERS   29
    Case Studies: The women's dam, Burkina Faso   35
                   Water for health in Kenya   37
                   Water for Rochina, Brazil   39
                   Canal hurts Colombian women   40

4. WOMEN AND FORESTS: FUEL, FOOD AND
    FODDER   42
    Case Studies: The story of Gadkharkh village, India   57
                   The Ghorepani Project, Nepal   60
                   Women in forestry, Kenya   63

5. WOMEN'S ENERGY CRISIS   66
    Case Studies: Afi : a woman from Ghana   80
                   The life of Ione Halley, Guyana   83
                   Mombamba Women's Group, Kenya   85

6. HUMAN SETTLEMENTS: WOMEN'S ENVIRONMENT
   OF POVERTY   87
   Case Studies: A child of Delhi   97
                   The women of Bhopal   98
                   Invisible women: purdah in Pakistan   98
                   Squatter upgrading, Lusaka, Zambia   101
                   Women's Construction Collective, Jamaica   104
                   Baldia Soakpit Project, Karachi, Pakistan   105
                   Urban agriculture, Lusaka, Zambia   108

PART II   WOMEN AND ENVIRONMENTAL CONSERVATION

7. WOMEN WORKING FOR CONSERVATION   113
   Interviews: Vandana Shiva, India   117
                   María José Guazzelli, Brazil   120
                   Shimwaayi Muntemba, Zambia   121

8. TRAINING WOMEN   123
   Case Studies: Training and technology in Senegal   126
                   INSTRAW   128
                   Pakistani women visit India   129

9. PLANNING THE FAMILY: A WOMAN'S CHOICE?   130
   Case Study: Planned parenthood and women's
                   development   136

10. WOMEN ORGANIZE THEMSELVES   140
    Case Studies: Green Belt Movement, Kenya   147
                   Ação Democrática Feminina Gaúcha,
                   Brazil   148
                   Women promote appropriate technology,
                   Guyana   149
                   ORAP, Zimbabwe   150

11. THE INTERNATIONAL RESPONSE   153
    Case Study: Luangwa Project, Zambia   170

12. WORKING TOGETHER FOR THE FUTURE   171

    References   180
    Abbreviations   195
    Resource Organizations   197
    Contributors   200
    Index   204

DEDICATION

JOAN DAVIDSON, 1943–1992

To Joan

I still cannot believe that you are far away now. I thought you would always be there: when there was an international meeting on women's roles in environmental management; when we had to lobby environmental organisations to look at gender aspects; when an article had to be written, or a book. . . .

It is already more than seven years ago that we began to plan 'our book'. I first met you at an international conference, and I was amazed to see a young woman telling the participants that they could not keep women and gender aspects out of their conservation work. Together with other women, we formed a working group to promote women's voices on environment and development. After the UN Women's Conference in Nairobi (1985), a plan to bring together in one book the environmental experiences of women around the world was formed. I was looking for a co-author, and thought of you; a journalist, environmental expert and fighter for women's rights.

From that time on, you have been not only a colleague, but also a very dear friend. We met many times, mostly in London, but we also travelled together to the IUCN meeting in Costa Rica and to the African Women's Assembly on Environment in Zimbabwe. You always paid warm attention to me and to others, bringing your wonderful smile, and small gifts reflecting our friendship. You were constructively critical of our texts and strategies, and showed enormous dedication to the improvement of the position of women, particularly in the South. You were an artist, and gave that artistic touch to your texts and your selection of illustrations. For you, sustainable development began close to home, as your book *How Green is Your City?* exemplifies.

You continued to work in the field of women and environment, developing the concept of primary environmental care and working as an environmental advisor to Oxfam. It is ironic that you called your last book, about poverty in the global environment, *No Time To Waste*. Your last mission was your participation in the UK delegation to the UN Conference on Economic Development in Rio, in 1991.

You have always worked very hard, driven by the desire to contribute to a better world for all people. Your husband, John, and your sons Ben and Dan have always supported you. You were a pioneer, a thinker, a worker, a friend. And now you are far away. But you'll always be in our work and in our hearts.

Irene Dankelman
Malden, January 1993

AUTHORS' PREFACE

It has not been possible, in this book, to convey adequately the drudgery and the suffering so many Third World women must face in their daily struggle to survive and care for their families. Nor have we done justice to the extraordinary resilience and energy these women display in impoverished and sometimes dangerous environments. Northern women, writing about life in the South, can do little more than try to give some voice to the voiceless.

This book is the product of women working together. Many busy people, a majority from the Third World, have described their direct experience of women's problems, and the continuing search for ways to solve them. We are indebted to all of them for their unstinting willingness to participate, their lively contributions, and their tolerance. We hope they will forgive such errors and omissions as have crept into the text.

The main contributors are listed at the end of the book. We are grateful to many of them for commenting on earlier versions of the text. We would like to thank especially the Honourable Victoria Chitepo for her support, Ann Clark and Sylvia Kreyenbroek for their substantial work on Chapters 2 and 3, and Frances Dennis of IPPF, Joan Martin-Brown of UNEP, Sue Milner of OXFAM UK and Dorothy Myers for their help on other sections.

The Dutch Ministry of Foreign Affairs gave financial and moral support for the preparation of the book. We are also indebted to the Netherlands IUCN Committee, Stephanie Flanders, Liz Hopkins, Perdita Huston and Mark Halle at IUCN in Switzerland, and to Margaret Busby, Sara Dunn and Sarah Stewart at Earthscan for their invaluable assistance on editing and the production of various drafts.

Finally, we want to thank our families and friends for all their support, and the women of the South who have inspired us.

Irene Dankelman and Joan Davidson
London, November 1987

FOREWORD

Ninety per cent of Third World women depend on the land for their survival. They are the world's farmers: they grow the crops, gather the firewood, tend the animals, bring in the water. In my own country of Zimbabwe, women in most of the rural areas are the main producers – growing food not only for their families but for sale. Often, they must do this alone, for their husbands and sons have left home for the cities or the mines. Their traditional farming methods, created of necessity to ensure future harvests, and passed down through generations, help to sustain the environment in which they live and bring up their families.

I know the work these women do – it is arduous and neglected. International agencies and governments have everywhere ignored the vital part that women play in caring for the environment. Their voice, like their knowledge and experience, is simply not heard. My own government is beginning to recognize the position, and the need to advance the status of women. Their participation in environment and development programmes is now a priority. Internationally as well, thanks to organizations such as UNEP, FAO and IUCN, awareness of women's crucial role in the environment is growing.

This book builds upon these efforts and advances the argument for listening to what women have to say. It describes not only how women become the victims of environmental crisis, but how they are responding. All over the world, women are taking action against the destruction of natural resources on which their lives depend.

One book cannot suggest all the solutions. What it can do – and does – is provide a valuable starting point, where not just crises, but examples of positive action, are examined and women's environmental experience is treated with the respect it deserves.

All too often, development agencies are accused of implementing projects with little concern for the real problems people face and scant regard for their social and cultural circumstances. This book helps us to understand these complex issues and adds tremendous impetus to our efforts to link improvements in women's emancipation with the care of the earth.

The Honourable Victoria Chitepo
Minister of Natural Resources and Tourism, Zimbabwe
Harare, November 1987

INTRODUCTION:
WOMEN HAVE SOMETHING TO SAY

Most rural women are directly dependent on their immediate environment, and their own skills in using it, for the daily necessities of life (Rocheleau, 1985:8).

LINK TO THE ENVIRONMENT

Women, particularly those living in the rural areas of Third World countries, play a major role in managing natural resources – soil, water, forests and energy. Their tasks in agriculture and animal husbandry as well as in the household make them the daily managers of the living environment. They have a profound knowledge of the plants, animals and ecological processes around them.

Women also participate in the commercial sectors of society and the raw materials they use in rural enterprises are vulnerable to environmental degradation and contamination. As farmers and traders, then, women experience environmental problems as directly undermining the basis of their daily lives.

Environmental degradation

A healthy environment is crucial in order to meet the primary needs of the world population. Yet, as the World Conservation strategy, the 1986 World Resources report and the report of the World Commission on Environment and Development (1987) show, the scale of environmental destruction is alarming:

- If the present rate of land degradation continues, almost one-fifth of the world's arable land will have disappeared by the year 2000 (IUCN, 1980).
- Each year six million hectares are degraded to desert-like conditions. Over three decades, this would amount to an area as large as Saudi Arabia (WCED, 1987).
- Throughout the developing world, water polluted by sewage or

industrial wastes has serious consequences for human health; in India 70 per cent of all surface water is polluted (World Resources, 1986).

- More than 11 million hectares of tropical forests are destroyed every year. Over 30 years, this would amount to an area about the size of India (WCED, 1987).
- Each day an average of one animal or plant species becomes extinct; most of these will disappear without ever having been discovered (IUCN, 1980).

Together with acidification of the environment, the rising level of carbon dioxide and destruction of the ozone layer in the atmosphere, these developments pose a profound threat to the present and future environment.

The poorest are the hardest hit

The principal victims of environmental degradation are the most underprivileged people, and the majority of these are women (Senghor, 1985). Their problems, and those of the environment, are very much interrelated. Both are marginalized by existing development policies. And because of the complex cycles of poverty, inappropriate development and environmental degradation, poor people have been forced into ways of living which induce further destruction (Aidoo, 1985). Third World women often have no choice but to exploit natural resources in order to survive, even though they may have the knowledge to promote sustainability.

The links between poverty and the state of the environment have only recently begun to be recognized by environmentalists, development specialists and those engaged in raising the status of women. In the international agencies, a number of significant events reflect this increased awareness:

- the highlighting of the issue of women and environment by the conference held at the close of the UN Decade for Women in 1985;
- "Women Nurture the World", a series of workshops organized by the Environment Liaison Centre at the NGO Forum '85 in Nairobi;
- the establishment of the Committee of Senior Women Advisers on Sustainable Development by the United Nations Environment Programme (UNEP) in 1986;

- a caucus on "Women, Environment and Sustainable Development" at the international conference on Conservation and Development – implementing the World Conservation Strategy, sponsored by IUCN and held in Ottawa, Canada, in 1986;
- the establishment of a Working Group on Women, Environment and Sustainable Development under the aegis of IUCN in 1987.

More important, in practical terms, has been the growth of environmental action by women's groups around the world. The Chipko movement in India and Kenya's Green Belt Movement are just two examples of many new initiatives designed to help women tackle the problems of environmental devastation. Many of them are described in later chapters.

THIS BOOK

The project on which this book is based was financed by the Netherlands Ministry of Foreign Affairs and carried out under the auspices of the Netherlands IUCN Committee. The aim of the volume is threefold: first, to examine the relationships between women and their natural surroundings; secondly, to show how women deal with the environmental crises they face; and thirdly, to look at the response of international and government agencies.

The first part of the book explores women's involvement in the use and management of natural resources – their role in agriculture, water supply, forestry, the collection and use of fuelwood and the development of other energy sources. The special problems women face in human settlements are also discussed here. Examples and case studies illustrate the effects of environmental degradation on women and the initiatives they take.

In the second part of the book, the position of women in environmental conservation is examined, with an emphasis upon their practical activities and their role in education and training, family planning and local organizations. The activities and policies of international agencies are described and a final chapter sums up the picture and presents a strategy for action.

To restore and conserve the environment, a worldwide reorientation of development towards sustainability is needed at all levels of society – from the grassroots to international action. Women are among the most important and best experienced actors in bringing about that sustainability.

We sing about our pain and our suffering.
It is a bad dream and no one understands.
One day our life will be different.

–Song of Ethiopian women in a Somalian refugee camp

If there must be a war, let the weapons be your healing hands, the hands of the world's women in defence of the environment. Let your call to battle be a song for the earth.

–Mostafa Tolba, Director, United Nations Environment Programme, Nairobi 1985

WOMEN, ENVIRONMENT AND NATURAL RESOURCES

PART I
WOMEN, ENVIRONMENT AND NATURAL RESOURCES

CHAPTER ONE
Why Women?

The role of women is crucial in meeting the big crises of the world today (Fresco, 1985:24).

It is difficult to talk about women as a whole without ignoring the vast economic, cultural and social differences between them. Even if we were to consider only women in the Third World, we would find it difficult to generalize. The lives of women in India, for instance, are different from those in Ghana or Peru; and within countries, similar gaps in income and culture exist.

Nevertheless, it is possible to outline the general shape of women's living conditions in the rural areas of the Third World. They share, first of all, their poverty: roughly 75 per cent of the world's population are among the poorest, and women make up the majority of the poor. Secondly, wherever they live, they are bound together by the common fact of their tremendous work burden. Time-budget studies show that women not only perform physically heavier work, but also work longer hours than men. In Tanzania women work an average of 3,069 hours per year; men work an average of 1,829 hours (Taylor *et al.*, 1985). In Bangladesh it has been estimated that women spend an average of 10–14 hours per day on productive labour. Household tasks are not taken into account in this estimation.

An Indian agricultural worker's day is typical. She rises at 4 a.m. She cleans the house, washes clothes, prepares the meal for her husband and children, and leaves for the fields at 8 a.m. She works there until 6 p.m., in the meanwhile nursing the small children she took with her. On her way back home she collects fuelwood, and, if necessary, drinking water. She cooks the evening meal, cares for the children and tends to the animals. At about 10 p.m. she goes to bed. On such a day she might earn less than two rupees.

Like this Indian woman, rural women have traditionally been the invisible workforce, the unacknowledged backbone of the family economy. Men go out to work, enter into commerce, make the decisions in the village and household, and are more likely to be chosen as spokespersons when it comes to dealing with government or development agencies. But when we look more closely at the types of work that women do, we can distinguish three main areas, all crucial to keeping the family and indeed the rural economy alive. These three areas are: survival tasks, work in the household, and income generation.

SURVIVAL TASKS

Survival tasks are those essential for daily life, and it is for these that women are largely responsible. They grow the food crops, provide water, gather fuel and perform most of the other work which sustains the family. A 1985 United Nations Food and Agriculture Organization study indicates that in some Third World regions, especially in Africa, women account for 80 per cent of agricultural production. And women's involvement in food-growing is increasing steadily. In Malawi in 1966, women accounted for less than 70 per cent of the agricultural labour force. By 1972 this figure had risen to 90 per cent.

A certain division of labour is evident in the agricultural sector. Women are generally responsible for sowing, weeding, crop maintenance and harvesting, as long as these tasks have not been mechanized. Men, on the other hand, look after field preparation. Subsistence agriculture – the growing of food crops – is almost exclusively a women's task. But women's participation is increasing in cash-cropping as well. Care of small animals is also often their responsibility.

The supply of water – vital for the survival and health of the family and for farming – is exclusively the concern of women and children. A study in East Africa showed that carrying water can use 12 per cent of the calorie intake of women, and, in dry and steeper areas, up to 27 per cent. In a village close to New Delhi, water collection takes an average of an hour each day. But as Chapter 3 of this book shows, collecting water can take a lot longer, and each load can weigh as much as 25 kilograms.

For their energy supplies, the rural areas of the Third World depend mainly on biomass such as fuelwood, crop residues and manure. Seventy-five per cent of rural energy supplies (and 90 per cent in Africa) come from biomass. Fuel collection, where it is not commercialized, is mainly a task for women, with children's help. Depending on the ecological characteristics of the area in which they live, women may spend up to five hours a day on fuel collection. The relationship between these survival tasks and the ecological situation is described in coming chapters.

HOUSEHOLD TASKS

Activities in the home are almost exclusively the responsibility of women, although older children may occasionally assist. These tasks return every day and absorb hours of time. Food preparation and cooking are a good example. In a study area in Peru, an average of four hours every day are spent on cooking (ILO, 1986). Yet women in many cultures are often the last in the family who may eat, and they take less than the other family members.

WOMEN'S INCOME

Throughout developing countries, women contribute substantially to the family budget through income-generating activities – food processing, trading of agricultural products and the production of handicrafts. This is particularly the case for the growing number of female-headed households where men have migrated to cities, mines, plantations, or abroad. A recent survey shows that the percentage of female-headed households south of the Sahara in Africa is already 22 per cent; in the Caribbean area 20 per cent; in India about 19 per cent; in the Far East and North Africa 16 per cent; and in Latin America 15 per cent. Locally much higher percentages can be found where women are the sole providers. Even where a woman is not completely alone, her contribution to the budget is of utmost importance to the family, the more so because women spend more of their income on family welfare.

It is clear that women fulfil a great number of essential tasks, yet they and their labour are often unrewarded. "Although women represent half of the world's population and one-third of the official labour force, they receive only one per cent of the world income and own less than one per cent of the world's property" (UN Conference, Copenhagen, 1980). Notwithstanding their important role, women have only very limited access to and control over income, credit, land, education, training and information.

Recent developments have worsened the position of women: Western colonization, the increasing dependency of Third World countries on a Western monetary economy, developments in technology such as agricultural modernization, the sharpening worldwide division of labour, and increasing religious fundamentalism have all brought extra problems for women. The accelerating degradation of the living environment is the latest and, in many ways, the most dangerous of the threats they face.

CHAPTER TWO

Land: Women at the Centre of the Food Crisis

Land, particularly healthy soil, is the foundation on which life depends. If the land is healthy, then agriculture and pasturage will yield food in plenty. If it is not, the ecosystem will show signs of strain and food production will become more difficult. Because women are at the centre of world food production – producing more than 80 per cent of the food in some countries – any analysis of land resources must include an appreciation of their central role.

LITTLE FOOD, NO LAND

The present world food situation is one of the great modern paradoxes: about 500 million people – the largest number ever – are seriously malnourished while world food production has reached the highest levels in history.

Hunger in the Third World is not necessary. Official Food and Agriculture Organization projections show that the earth can provide more than enough food not only for our present population of 5,000 million, but also for the 6,100 million people expected by the year 2000 (FAO, 1981). But these numerical calculations do not take account of the problems of food distribution, of economic control over food resources, and the politics of food dependency. Access to food is not simply a question of land availability, but of social, political and economic power. The poor have none of these.

Expanding cropland will offer only a limited solution to the landlessness of the poor. Success would depend upon good-quality land being available to those most in need. Of the world's 13,250 million hectares of land, about 30 per cent is estimated to be arable, half of which is now under cultivation. Just over half of the land presently under cultivation is found in developing countries, but it is inhabited by

Figure 1: *The landless and near-landless as a percentage of total rural households*

Bolivia	85%	Ecuador	75%
Guatemala	85%	Peru	75%
Indonesia (Java, 1971)	85%	Brazil	70%
El Salvador	80%	Dominican Republic	68%
Philippines (1972)	78%	Colombia	66%
Sri Lanka (1970)	77%	Mexico	60%
Bangladesh (1973)	75%	Costa Rica	55%
		India	53%

Source: Sinha, 1984.

three-quarters of the world's population. Within countries, there is severe inequity of land ownership (see figure 1).

Even with the initiation of agrarian reforms, the politics of land ownership often work to ensure that the most productive land remains in the hands of a few. Where political power resides with a land-owning élite, governments allow private estates to expand further and protect their boundaries. In Colombia, land reform has meant the modernization and capitalization of existing estates, leading to even greater concentration of ownership (Leon de Leal, 1985). In El Salvador, the land reform of 1980 brought no significant change in the plight of the country's 2.5 million landless or near-landless peasants (Pearce, 1986). Here, as in many Third World countries, the poorest are squeezed on to marginal lands which are steep, infertile, dry, subject to pests or disease or covered with rainforest. Their attempts to grow subsistence crops result only in increased erosion and the destruction of water and fuelwood resources. Where many people are forced on to poor land, fallow periods diminish or disappear and the possibilities for soil recovery are reduced. Scarce energy sources cause people to burn manure and crop residues to meet their fuel needs, and the loss of these traditional sources of soil nutrients decreases land fertility. It is estimated that the annual burning of 400 million tonnes of dung depresses the world's grain harvest by over fourteen million tonnes (Spears, 1978).

Women produce food
Many of those dispossessed of land by the increasing concentration of ownership are women and their children. Women have title to only one

percent of the world's land. Yet they produce more than half of the world's food – and in countries of food scarcity the percentage is even higher. Women produce more than 80 per cent of the food for sub-Saharan Africa, 50–60 per cent of Asia's food, 46 per cent in the Caribbean, 31 per cent in North Africa and the Middle East and more than 30 per cent in Latin America (FAO, 1984; Foster, 1986).

Women make up the majority of subsistence farmers. In most rural cultures, it is their work which provides a family with its basic diet and with any supplementary food that may be obtained from barter or from selling surplus goods. Underestimating the amount of agricultural work done by women is very common, for statistics most often measure wage labour, not unpaid kitchen garden work. Moreover, in some cultures men do not wish to admit that their wives, mothers and daughters do agricultural work. For these reasons, the vital contribution that women make to food production is consistently under-represented (Taylor *et al.*, 1985).

Women also participate actively in cash-crop production, either as extra hands at harvest time or as employees on large farms. Women can sometimes spend more time in export production than men. In Nigeria, for example, women work more than men in the cocoa plantations, in coffee production for export, and in national market crops such as rice, grain, maize and cassava (Fresco, 1985). In Nicaragua in 1980, women made up 28 per cent of coffee harvesters and 32 per cent of cotton workers (IFDP, 1980).

MORE TECHNOLOGY, LESS SOIL FERTILITY

Increasing agricultural industrialization, particularly under the "Green Revolution" introduced in the 1950s, has had an enormous effect on women. This policy of intensifying food production through developing hybrid, high-yield seed varieties demanded extensive

Figure 2: *Percentage of total agricultural production by women*

Nepal	98%	Iraq	41%
Zaire	64%	Brazil	32%
Korea	51%	Colombia	20%

Source: Jiggins, 1984.

irrigation and increased mechanization, as well as the use of fertilizers and pesticides. Chemical and biological technology was applied on a large scale in South-East Asia, India, China and Mexico, with the goal of increasing food production. Yields did increase, often dramatically. Between 1974 and 1983 they grew at more than two per cent each year – largely as a result of the increased productivity of land already under cultivation. But in spite of this increase the people of South-East Asia and India are still among the least well-nourished in the world. The Green Revolution has produced no increase in per capita food consumption and has in many cases reduced it (Lappé and Collins, 1986). Instead, the Green Revolution has contributed to erosion, desertification, and greater concentrations of land ownership, removing land from those most in need.

Erosion and desertification are not merely a result of rainfall:

> Only appropriate land use can keep arid zones ecologically stable and biologically productive. Inappropriate land use can destabilize even humid regions, undermining biological productivity and causing desertification. Since the large majority of people in countries like India have livelihoods based on land, the long-term decline of the biological productivity of land undermines livelihoods and results in underdevelopment (Bandyopadhyay and Shiva, 1986:1).

Large-scale, mechanized, highly-technological agriculture is extremely taxing on land fertility. Widespread irrigation – probably the most effective way to increase yields – can cause waterlogging, a reduction of essential minerals, and salinization because of increased evaporation. More than a third of all land under irrigation is subject to salinization, alkalinization and waterlogging (UNEP, 1982). In some areas, 80 per cent of the irrigated land has been destroyed in these ways. Worldwide, salinization alone may require the abandonment of as much land as is now under irrigation (World Resources, 1986).

Synthetic petroleum-based fertilizers are also the cause of serious soil and water pollution. Experiments show that this kind of agriculture affects the metabolic balance in plants, leaving them more vulnerable to attack from pests and diseases. Farmers are then caught in the vicious circle of increasingly intensive (and costly) use of pesticides which in turn causes greater pest infestation (Guazzelli, 1985). There is growing concern over the developing immunity of many pest species as pesticide use accelerates, especially in developing countries.

Pesticide use also has serious consequences for health. During informal consultations in 1985, the World Health Organization estimated that one million cases of pesticide poisoning occur annually.

Not only human beings, but many other non-target species (and livelihoods based on them) are affected by pesticide poisoning, including livestock, fish, birds and bees (Bull, 1982).

The introduction of laboratory-bred hybrid crop species has also had negative consequences. In the district of Dharwar, India, for example, a mix of indigenous varieties of crops were cultivated, giving high yields of fodder, pulses and oilseeds. These varieties were drought-resistant and, in normal rainfall years, produced food that could be stored for the drought years. The introduction of a single hybrid Jower (sorghum) not only reduced fodder production, but also made the crop susceptible to failure in short drought periods because of a decrease in absorbent organic matter being added to the soil (Bandyopadhyay and Shiva, 1986).

Promoting the use of hybrid seeds also diminishes the genetic resource base of many crops, leaving farmers dependent on fewer, perhaps less adaptable, varieties in the face of changing weather conditions. A Sri Lankan farmer remembered 123 varieties of red rice; now only four remain. He has one acre of paddy which, in a very good year, produces 100 bushels of rice – a surplus of 25 bushels. But, because each of the traditional varieties was less vulnerable to severe conditions than the hybrid variety now in use and because the hybrid does not keep long in storage, he requires an ever-increasing surplus on which to live (ICDA, 1985). In this way, new technology has reduced farmers' self-reliance.

The pressure to engage in industrialized agriculture has increased with the changing global pattern of food trade. Many developed countries depend for their food supply on land use in developing countries. United States hamburgers are increasingly made from the beef produced on cattle ranches in Latin America. Extensive ranching, which has reached into the Amazon basin, is now destroying a reservoir of plant and animal species of great genetic diversity (Lappé, Collins and Fowler, 1978). Recent studies have also shown that Dutch agriculture uses a land area in the Third World five times the size of its own cultivated land to supply the cattle fodder and raw materials for food products (Netherlands, 1986).

Patterns of agriculture that demand enormous economic and chemical inputs do nothing to ameliorate critical food shortages. Instead, they increase these shortages through dependence on an economic and industrial order which excludes the very people who are most in need.

More work for women

All of these consequences of industrialized agriculture impinge upon the everyday lives of women attempting to glean food, fuel and water from their environment. Many effects are confined to women because of the gender-specific division of labour in rural agricultural societies. In most cultures, women are responsible for food processing, fuel and water gathering. "In the agricultural sector of Africa, women perform 80 per cent of the storing, 90 per cent of food processing, 60 per cent of the marketing and 50 per cent of the domestic animal care ... often with few, if any, modern tools" (ILO/INSTRAW, 1985). However, specific tasks allocated by gender are renegotiated in response to changing conditions – in ecological conditions, in social structures, and in economic relations, among many others (Creevey, 1986). Thus women's agricultural work – in fields, weeding, harvesting, irrigating, tending poultry and small grazing animals or cattle, food preparation and marketing – depends on where they live and their place in the rural economy (ISIS, 1983). For example, husbandry of livestock is often women's work, but what that actually means will vary. In Pakistan, women care for small livestock and cattle. In Chile, 80 per cent of the small-scale women farmers also tend livestock. Up to 90 per cent of rural women in North Africa, Asia and Latin America keep poultry, and in several Sahelian countries women are the major goat tenders. Whether this responsibility includes carrying water, gathering fodder, securing animals from predators, preparing for market or marketing will be different in each country.

The spread of erosion and desertification makes all of these tasks more difficult or even impossible. As women's contribution to their own well-being becomes more problematic, their marginalization increases. When access to fertile land diminishes – as intensive agriculture eats up the small plots of the poor – women are displaced to more distant, fragile and less fertile lands, or even become landless labourers. Without land, women have no access to credit, training or other aids to production.

The number of landless people worldwide is growing dramatically (UNEP, 1980). In Pakistan, India and Bangladesh, one-quarter of the population has no land and this group is growing fast. The majority of the landless are women. Where efforts are made to address the problem of landlessness, as with land reform, the situation of women is exacerbated: women's traditional and existing land rights do not figure in the calculation as to how to reallocate land (Foster, 1986).

With the displacement of subsistence agriculture to less fertile areas, women who still have access to land must travel longer distances to their fields. When they arrive, they must work harder to compensate for severe erosion and low fertility. When their plots are exhausted, they must wander elsewhere in search of food. In the Ivory Coast, women have had to leave their own fields because of expanding agro-industrial coconut and palm oil plantations. They were forced to move into the Tai Forest, where farming not only caused environmental damage, but the land they used was not suitable for permanent agriculture (Bamba, 1985). Environmental degradation of this kind has other consequences as well. In Mozambique, not only political instability but erosion and environmental degradation are causing increasing numbers of refugees to move into Zimbabwe, imposing extra burdens on the capacity of Zimbabwean resources.

The "double" day

As the transition to cash-cropping makes inroads on rural agriculture, women's activities become both more burdensome and less socially valued. Marginalization occurs where there is a desperate and pervasive need for cash to meet family needs. If those needs can only be supplied by the market, unpaid traditional roles will no longer evoke respect, thereby undercutting the authority of women (Huston, 1985). Women's subsistence agriculture suffers from neglect where large-scale agriculture absorbs labour, land and economic resources. At the same time the labour necessary for survival increases. In Cameroon, for example, men were given land, water, seeds and technical training to enable them to produce rice for sale. Women were then expected to carry out their traditional agricultural tasks in the cash-crop rice fields, as well as cultivating sorghum for their families (Foster, 1986). This pattern in women's lives, called "the double day", has been a perennial accompaniment to transitions from rural to industrial capitalist economies. And the work that women are expected to do in growing cash-crops – planting and weeding – is often more gruelling than men's work, for men generally run the farm machinery. The increase in yields that follows the use of fertilizers, pesticides, high-yield seed varieties and mechanization can mean more work for women. In Sierra Leone the introduction of tractors and modern ploughs resulted in a decrease of the working day for men in the rice culture, but women had to work 50 per cent more to finish weeding and maintaining the larger fields.

The use of pesticides can also displace women. In Kenya and Uganda, the widespread use of insecticides and fungicides has made many women, who formerly did 85 per cent of the weeding by hand, now redundant (Morse, 1978). And finally, because large-scale agriculture attracts migrants from other areas in search of work, it increases the competition for jobs.

These changes have meant that the proportion of women among farmers in India has decreased from 45 per cent in 1951 to 30 per cent in 1971, while the participation of female labourers rose from 31 to 51 per cent over the same period (after the introduction of the Green Revolution). For these women, the high-yielding rice technology may have increased employment opportunities, but particularly for women of poorer households, the result was also an increased workload without any improvement in their standard of living (Agarwal, 1986).

In other ways, too, large-scale industrialized agriculture affects women for the worse. Intensive fertilizer use, for example, requires a great deal of water. But in many countries there is no surplus. In Zimbabwe and India the water table is steadily falling as a result of irrigation and fertilizer use. This may also contaminate local water supplies. In Zimbabwe, women who were used to collecting drinking water from wells in the fields where they worked can no longer do so because the well water is contaminated by fertilizers used on the land (Nyoni, 1985). And in Haiti, women use the water from the irrigation ditches since they do not have access to a source free of chemical runoff (personal communication, Susan Quinlan, IFDP, 1986). Thus, where once women could fulfil their family needs for water locally, they now must travel several kilometres or use contaminated sources.

In the state of Rajasthan, India, now on the brink of a desertification disaster, wells and once-flowing rivers are dry. In 1975, the World Bank and its partners introduced capital-intensive irrigated farming in that area. A region of just 60 centimetres of rainfall a year was planted with sugarcane, which has such a high water requirement that groundwater levels have fallen dramatically. Crops fail and people starve (Shiva, 1985–2).

As food providers, women play a central role in the nutritional intake of the family. But this is similarly affected by the development of large-scale monocultures producing crops for export at the expense of subsistence foods. The complementary structure of rural diet – beans and corn or lentils and rice, for example – has been undermined not only by these shifts but also by intensive marketing of commercially processed products such as Coca Cola and bread (Lappé and Collins, 1978; ISIS, 1983). Agricultural change linked to glistening images of progress alters nutritional intake, and makes women's task of feeding the family that much more difficult. Health suffers and malnutrition increases.

Finally, large-scale agricultural expansion places impossible demands on women, who may have access to land but rarely to the capital or credit to invest in machinery, hybrid seeds or chemicals. The inputs required by Green Revolution agriculture are usually beyond their economic reach. "When you cannot afford to drink the milk from your own buffalo, but must sell it to buy wheat, what possibility is there of purchasing fertilizers and imported seeds?" (Taylor *et al.*, 1985). In some villages in India, women have pawned their silver jewellery, their only property, to meet the costs of new seeds, fertilizers and pesticides, thereby removing any economic independence they might have had (Shiva, 1985–1). But for most women who must work all day in the fields raising wheat or cassava, any investment in agricultural machinery is far beyond their reach, culturally as well as financially (Whitehead, 1985). To an Indian woman whose wages may be controlled by her husband and who must sit on the floor while he sits on the only chair, owning or even driving a tractor is inconceivable. And yet some development projects are blind to these circumstances, assuming that the only difficulty for women is a lack of funds. Because of the centrality of women's labour to rural survival, lack of attention to the particular effects that the development of industrial agriculture has had on women is especially damaging. The responsibilities and skills of rural women in fulfilling their families' food, fuel and water needs must be accorded their full economic and social significance if development efforts are to be of real assistance to Third World people.

WOMEN LOSE THEIR RIGHTS

Most developing countries have a long history of colonial rule which imposed laws and social structures particularly harmful to women. Among these are inheritance laws, legislation on land ownership and transfer, and social restrictions on women which seriously limit their activities and aspirations. These patterns have occurred extensively in Africa, but European laws and customs have also altered the place of women in Central America so as to reduce their power. Inheritance laws and communal rights to land which once allowed access by women have been replaced by title-deed systems which, by law or custom, restrict land ownership to individual men (Poldermans, 1985; Jiggins, 1984). Spanish colonization in Latin America brought with it an ideology of chastity and dependence which has dominated women's lives. In Africa, where market organization of women traders once allowed women political and economic power, and where sister/brother inheritance and kin cooperation patterns allowed women an alternative to dependence on a husband, European patriarchy undermined both (Sachs, 1982).

Men migrate

These same developments, especially since the Second World War, have fostered the growth of industrial centres which draw men away from rural communities, removing their labour from subsistence farming. The increasing migration of men to the cities, to mines, to export agriculture, or to work abroad has caused the number of female-headed households to rise dramatically. This often means that the entire responsibility for feeding, clothing and housing children rests on the woman's shoulders. Recent studies of 74 developing countries show that women already head more than a fifth of the households in Africa and the Caribbean, and 15 per cent of those in Latin America, the Middle East and North Africa. For some countries, the figures are much higher: in Kenya, Botswana, Ghana and Sierra Leone almost half the households are headed by women (Taylor *et al.*, 1985). In Lesotho they are called "widows of the gold", as their husbands migrate seasonally to South Africa to work in the mines, where eight thousand have died in the past decade. A US Agency for International Development (USAID) study shows that female-headed households are the poorest group in every country (Foster, 1986).

Discrimination in wages and training

When rural women do participate in the wage labour economy, they face discrimination and lower wage rates. In countries which are trying to meet rural needs through the development of agro-industry, discrimination in job classification and wages is especially intense. Helen Safa reports that 70 per cent of jobs in food processing for export are women's jobs, because they can be paid at much lower rates. In countries like Mexico, which has the highest investment in food processing of any country in the world, agro-industrial development is seen as a solution to rural poverty. In these industries the lowest paid jobs are reserved for women. Strawberry-processing plants in Zamora employ 10,000 young rural women to stem strawberries, select and pack them at wages well below the legal minimum (Arizipe and Aranda, 1986). In the Philippines, women working on sugar haciendas earn 2–3 pesos to men's 4–5 (Eviota, 1986). More than half of the working women in Java, Indonesia earn less than 3,000 rupiahs per year, while only 14 per cent of the men earn so little (Taylor *et al.*, 1985).

Where women remain in subsistence agriculture their central position is usually ignored, even by development professionals. Thus training and agricultural extension programmes often fail to reach women. Out of a study of 46 African countries it became clear that less than 4 per cent of extension workers who advise women are themselves women. In Asia, the proportion is 7 per cent. The impact of even this small percentage is greater than it appears, because in Moslem cultures it is prohibited for women to be taught by a man (FAO, 1984). In some Hindu and Christian cultures, too, customs preclude an open learning atmosphere between women and a male non-family member.

But this is not the sole problem. The teaching of non-middle-class women is not even considered in many areas. "Women have not been part of the mainstream of educational activity anywhere in the developing world" (ISIS, 1983: 176). Yet illiterate women, out of reach of extension workers, are especially vulnerable to the injudicious use of dangerous chemicals. Spraying of fields and local storage of unsafe chemicals are special hazards for these women, whose children also suffer if their mothers are poisoned during pregnancy (Dankelman, 1985). In Central America, more and more women are found with poisoned milk from pesticides. Such consequences for women and children are wholly overlooked when the focus of attention in evaluating development is narrowed to the amount that yields have increased.

All of these unplanned effects of development – development that can be described as "gender-blind" – have led to increasingly desperate circumstances for women and their families. In Brazil, for example, they leave the countryside for the large cities, but the limited options for work there have intensified their poverty. Many women must work in the "informal" sector, selling food and other goods on the streets or doing domestic work. Others are forced into prostitution. And high rates of delinquency are observed among children (Guazzelli, 1985).

Listen to the women
Gender-blind development has another consequence: it undermines ecologically sound traditional agricultural knowledge. It should be clear that women, as the world's most important food producers, are directly dependent on a healthy environment. It is also becoming more and more obvious how much rural women themselves are conscious of this dependence and how much they know about their natural

environment, soil conditions and crops. As a rural woman from
Zimbabwe said: "My environment is the basis of my economy and my
total survival. It is from the land that I get my food" (Nyoni, 1985).

Women's agricultural methods, practised successfully for over forty
centuries in countries like India and China, adapt to the environment
and are sustainable without long-term damage to the land (Shiva,
1985–1). The knowledge and experience of generations permit women
to have great flexibility in cropping practices. For example, the serious
decline in soil fertility in many parts of Africa has caused them to shift
their cropping from maize to cassava. Although the traditionally-used
cassava root offers less nutritional value than maize, women have
begun to use all parts of the plant, including its leaves, in meal
preparation so that there is actually an increased intake of calories and
proteins (Fresco, 1985).

Women's agricultural knowledge provides security for themselves
and for others. As long as women are still engaged in seed selection, the
future survival of traditional crops is assured. In Zimbabwe, a woman
too poor to purchase new millet seeds used traditional ones. Later
droughts caused others' harvests to fail, but her crop survived. A
women's organization purchased 25 bags of her traditional seeds and
distributed them to other women throughout many villages (Van
Brakel, 1986).

Today's hybrid varieties do not reproduce fertile seeds; the farmer
must buy new seed every year. She is now completely at the mercy of the
seed supply system. Over time, she may lose her traditional knowledge
of seeds since she no longer selects them after each harvest for the
following year. Modern agricultural practices thus contribute to the
genetic erosion of crop varieties, and women become more dependent
on the purchase of hybrid seeds.

Women know that participating in the new agricultural technology
threatens their only means of control over their livelihoods: "In
Tanzania, when new hybrid maize seeds, fertilizer and pesticides were
allotted to men by the government, the women who do most of the field
work neglected the new crop. Although their workload with the old
crop was heavier, the profits from the new crop would, by tradition,
have gone only to men" (*WorldWIDE News*, 1986 (1):4). In Ghana, rural
women were reluctant to accept new hybrid maize seeds since the crop
had an unpleasant taste, was hard to prepare, was less resistant to
drought and insects, required different storage methods and depended
on chemical fertilizers which changed the taste. Although these
objections were rational they were not considered by the development
agency (Ahmed, 1985).

But the belief that women's knowledge cannot be scientific has kept it from being recognized and threatened its survival. In tribal Indian villages, women were growing high-yielding, indigenous varieties of rice, but because the women were considered backward and not scientific enough, "modern" agriculture was introduced (Shiva, 1985-1). There are, moreover, economic pressures against building on traditional knowledge – for those who profit by selling the prerequisites of industrial agriculture want to do so frequently. Indigenous methods do not require financial investment at every planting.

WAYS FORWARD

Because of this [food] crisis, we are interested in sustainable agriculture, not for luxury, not for economic reasons, but first and foremost for our own survival (Nyoni, 1985:53).

Sustainable agriculture is food production which respects both the natural and social environment. It is based on wise use of natural and renewable resources with moderate exploitation (Shiva, 1985-1). And sustainable agriculture is controlled by the community it supports, so that it may flexibly respond to the needs both of people and their environment.

In practice, sustainable agriculture requires:

- equitable access by all farmers to fertile lands, credit and agricultural information;
- the maintenance and support of independent agriculture over which rural farmers, both women and men, have control themselves;
- the development of cultivation, food processing and food storage methods which ease the intense demands on women's labour;
- a high degree of species diversification to maintain flexible cropping patterns;
- the conservation of fertile soils in which organic matter is recycled (to avoid dependence on imported nutrients); and
- an appropriate use of water and fuel resources.

A variety of methods is now in use to conserve land resources and the diversity of species. New styles of landscape planning encourage a diversity of habitats, integrating trees with crops. Mulching, multiple cropping and crop rotation, the use of compost and green manure, can all improve soil fertility along with the integration of animal husbandry, use of natural fertilizers, special varieties and pioneer plants for soils. Selective weeding and integrated pest control have improved crop protection. These methods have produced higher yields over a longer period and reduced the dependency on expensive external inputs. Furthermore, they allow agriculturists to maintain a healthy environment. They incorporate traditional knowledge and expertise, particularly of the woman farmer, increasing her self-reliance and her control over the process of agricultural production. Although these management methods for sustainable agriculture will not remove the injustices of land distribution or other inequalities and discriminations that characterize global patterns of food production, they are a start.

Locally, groups of women in diverse communities have been working towards wider goals. In Zimbabwe, women are demanding land ownership as well as participation in cooperatives (Weiss, 1986). In Nicaragua, where hard-won land reform makes access to land legally possible for women, the national women's organization is working for social changes that will improve this access, and has negotiated with unions for day-care facilities in agricultural areas (Davies, 1983). And Carmen Diana Deere reports that women in Nicaragua have pushed for communal eating areas at rural work centres, thus mitigating the "double day" (Feminist Theory, 49). In the Dominican Republic, local women's groups spread techniques of intensive small-scale gardening which were both agriculturally self-sustaining and soil-conserving on over-logged hillsides (Chaney, 1985). By January 1983, there were 6,000 of these household gardens in the Dominican Republic (Nash and Safa, 1986).

The Wum Area Oxen Project, begun in 1976 by the Cameroon government and the West German Agency for Technical Cooperation, is now transforming women's lives in the mountainous northwest of Cameroon. The use of oxen for ploughing, harrowing, raking, mulching and transport has replaced back-breaking hand cultivation which previously took from dawn to dark, and required leaving children unattended and household chores undone. The project offers farmers a two-month training course in the use and maintenance of draught animals. Preference is given to women's groups which can send at least four members to be trained; there are now 34 groups participating. At the start of the training session, every farmer or group is given a pair of oxen.

The use of oxen has increased the land under cultivation, and increased the number of crops harvested in a year. In many areas wheat, rice and soybeans are now planted in addition to the traditional groundnuts and maize. Bananas, plantains, cassava and yams are also grown and rotated with fodder crops and legumes to maintain soil fertility. Mixed farming can thus make even a small farm viable, and help ensure adequate long-term food supply (Barry, 1986).

Women in Gambia have embarked on a soil-preservation scheme by building dikes which prevent salt water from encroaching in their rice fields. The Gambian National Women's Bureau has also introduced orchard woodlots managed by village women's groups. These will bring firewood closer to home as well as improve nutrition by supplying families with fruits and vegetables. Women are planting nursery bed and spice plots as well. These practices enrich the soil, prevent erosion, increase crop diversity and provide fuel and shade.

In Rajasthan, India, where, as discussed earlier, large-scale sugarcane production has undermined local agriculture, women have been the driving force behind the Chipko movement. Acting against the advice of the Forestry Research Institute, women planted oak trees in deforested regions, so providing a new basis for water, fodder and fertilizer (Shiva, 1985–2). Elsewhere in India, the Self-Employed Women's Association (SEWA) of Ahmedabad has sponsored cattle and dairy projects for landless agricultural labourers. These women were not paid in cash, but in food grains or fodder; the switch to self-employment, and the development of dairies, milk cooperatives and a subsidized cattle-feed programme have given them income and independence. SEWA members now run the milk business themselves: they tend the cattle, sell the milk and deal with the banks and moneylenders. Says one observer: "SEWA is not just an office or an organization, it is a women's movement. It is definitely making a dent in the policies for poor women and their self-image" (Braun, 1984).

The following case studies describe some pioneering examples of sustainable agriculture from Africa, Latin America and the Middle East. They are a small sample of what can be done to improve the circumstances of women in agriculture. The International Fund for Agricultural Development (IFAD) and the UN Fund for Women (UNIFEM) are among the United Nations agencies which support these efforts. But much more remains to be done. Efforts will only succeed where the needs and desires of women, as well as the overwhelming responsibilities facing them, become the starting points for action. If underdevelopment is to be addressed, the farmer herself must be recognized and empowered. Enabling rural women to change their lives will require a restructuring of social and political controls over agriculture and land so that a woman may participate in the decisions affecting her life. This will be a long process, but progress toward this goal is not impossible, as the examples show. Development agencies, women's organizations and rural grassroots groups of all kinds must continue to use their energy and imagination to reclaim an agriculture that will sustain women, their families and the environment for years to come.

CASE STUDIES

WOMEN AND SOYBEANS IN TOGO, WEST AFRICA

According to Togolese tradition, women prepare the special condiment dawa dawa mustard sauce. Dawa dawa seeds are scarce and expensive; they take time to gather and require lengthy preparation. Deforestation in Togo has reduced an already limited supply of the trees from which the seed is collected and from which the wood, needed to boil the sauce for twenty-four hours, is gathered. The need to find an alternative to the dawa dawa seed has triggered a valuable development project for women.

In Togo, as in most other parts of Africa, women are responsible for feeding their families. Men provide the staple grains of sorghum or millet; women supply the vegetables, seasonings, meat and sauces. All are costly. The result is a frighteningly high level of malnutrition among Togolese children – up to 60 per cent in some villages during the "lean season". Both development and nutrition educational programmes have failed to solve the problem. Food aid is available through some local clinics, but it is often not suited to local tastes or physiology.

Since 1979, World Neighbors has collaborated with Family Health Advisory Services (FHAS) in Togo to help women look for resources near to hand as solutions to their problems. This approach has enabled them to make use of not only local crops but also their own skills in working cooperatively.

The soybean project

A major success has been the introduction of soybeans, a relatively new crop to West Africa, but one which has become the entry point for an integrated programme of farming and family health in the three countries of Togo, Ghana and Mali. Soybeans are an ideal crop for poor women: they improve both nutrition and the environment. With a 40 per cent protein content, these legumes fix nitrogen from the air and produce good yields without fertilizer – often double the yield of local beans. Soybeans grow fast and resist drought and many of the insect pests and parasites common in tropical West Africa. Apart from seed, no inputs are needed.

Soybean cultivation has been tried here before and failed. What is different now is the approach of Ayele Foly and Alice Iddi, the two West African leaders of FHAS. They looked first at women's immediate needs and then developed the project by helping the women to test a new idea on a small scale, encouraging them, and training local women to train others.

Even before the FHAS project, a few women were cultivating local soybean varieties; FHAS began to promote what these women had learned. The crop was first introduced as a legume that could be used to make sauces, not as a commercial crop in which men would have an interest. The men did not, therefore, resist when the women asked for small plots of land to grow the beans. FHAS recruited and trained a small number of volunteers to cultivate the crop on trial plots and to share their success with others. It then organized cooking demonstrations, showing women how to prepare their local dishes with soybeans, and getting them to compare the results with food cooked using their usual ingredients. Slides were shown of other villages where soybeans were used; exchange visits were arranged between soybean- and non-soybean-growing villages. These exchange visits and the workshops in women's homes have proved to be highly popular and have provided, FHAS argues, a more effective training environment than an unfamiliar urban study centre. The pioneers return from these workshops to their own villages, where they train other women to grow and use the soybeans. The process makes good use of women's natural leadership qualities and their willingness and ability to work together, and moves the project towards its long-term goal – to strengthen the status and economic position of Togolese women.

The key to gaining widespread acceptance of the new crop was the promotion of soybeans as a substitute for the seed bean used to make dawa dawa mustard, the highly prized ingredient of most local sauces. Later, women began to use the soybean in other dishes and to make a high-protein porridge for their children. FHAS helped mothers to identify malnourished children (by coloured armbands), and trained some of the women as volunteer health workers.

Benefits

In Togo, Ghana and Mali, more than a thousand women now cultivate soybeans as a result of the FHAS programme. Malnutrition among children has been significantly reduced. The projects have enabled women to generate an income for their families by selling surplus soybeans (raw or as flour or mustard) in local markets. They are saving money previously spent on imported flavourings or dawa dawa seeds. Now, women participate more fully in family decision-making – especially about the use of land.

Source: Peter Gubbels and Alice Iddi, 1986.

THE PINABETAL WOMEN'S ORGANIZATION, MEXICO

Living in the mountains to the south of the town of San Cristóbal de las Casas, Chiapas, are small communities of Indian peasant families, the Tzotzil. They have been there since the big landholdings and plantations ceased to be productive – a direct result of the overuse of land and the gradual transformation of feudal farms into agribusinesses. In the arid, stony areas which overlook the fertile valleys of the Grijalva River in central Chiapas, these families, and the women especially, are fighting the poverty which surrounds them.

There is a strong collective tradition in these communities. They have installed water points and electricity supplies, and built schools. But for women the story was familiar. However crucial they were to the success of the projects, women were allowed no part in the structure or decision-making of local organizations. The activities which directly concerned them, such as the rearing of small animals, were not given priority in the male-dominated cooperatives of most of the communities. Nor could women get loans to improve their economic position: there was consistent discrimination against them in the use of a local revolving loan fund.

In 1981, a group of women based in Pinabetal formed a collective to overcome this discrimination and to tackle some of the major problems the women face – lack of access to land, no roads, no schools, no work. After years of struggle, the Pinebetal Women's Organization has succeeded in buying land for the women to cultivate together.

The Pinabetal Organization now involves 30 peasant women. They work collectively to grow vegetables, raise sheep for wool and as a source of natural fertilizer, and are planning to keep dairy cattle to provide milk. Current projects, in part supported by OXFAM, include chicken rearing, water management, more vegetable plots and soil conservation measures.

As important as these achievements are, they are not the final answer. The Tzotzil women need to lift themselves out of their extreme poverty, but to accomplish this in the long run they need confidence, skills and experience. For this reason, the Pinebetal Organization runs study groups and evaluation and planning sessions, in which its members decide the shape and priorities of their collective. As Indian women, they also benefit from discussion about the situation of indigenous women farmers. The women of the Pinabetal Women's Organization now hope that their success will be replicated in neighbouring villages, where women want to find ways of working together to solve their problems.

Source: OXFAM, UK.

THE VACARIA PROJECT IN BRAZIL

Since the 1960s, Brazilian agricultural policy has emphasized Green Revolution technology as a model of production and development. That model, based on expensive foreign inputs (chemical fertilizers and pesticides, heavy machinery and hybrid seeds), has been promoted through subsidized bank credits. Production is characterized by monoculture and this has destroyed the peasant social structure in southern Brazil. Increasing impoverishment of the land is driving farmers into the Amazonian rain forests of northern Brazil and to the urban centres. In order to achieve social changes that would relieve the situation, Ação Democrática Feminina Gaúcha (ADFG), the Friends of the Earth in Brazil, is implementing a combined environment and development project coordinated by women.

With financial support from Friends of the Earth in Sweden, and the Swedish International Development Authority, a project on Low External Input Agriculture was begun in February 1985. The project involves the management of a farm and a training centre, with a programme for peasants, extension workers and students from agronomy and veterinary faculties. The administrative and technical coordinator for agricultural matters is a woman agriculturist, while a woman veterinary surgeon coordinates animal-related matters.

The farm is located at Vacaria in the highlands of southern Brazil. Its 50 hectares (of which 25 are natural woodland) are managed to include cropping, gardening, fruit growing, animal husbandry and agroforestry. The main goal of the project is to demonstrate that sustainable agriculture can make small farms viable, providing work for unskilled labour and slowing migration to the cities and rain forests. It is hoped that the demonstration will bring about changes in Brazil's agricultural policy.

Sustainable agriculture is based on the adoption of techniques that increase soil fertility or maintain fertility indefinitely. The approach aims to protect environment, keep the energy balance, and to control erosion without using chemical fertilizers and pesticides and by integrating a diversity of crops and animal production. Most of the inputs are locally or regionally produced so that food production is freed from international arrangements and trends that create dependency. The technology is also less expensive than modern technology – a distinct advantage in view of Brazil's current economic crisis.

At its inception, the ADFG project gained widespread support. But it was also strongly criticized: in addition to launching a new agricultural concept in opposition to large corporate pressure, two young women were running a project in a traditionally male domain. Since then, however results have been so successful that people from many different backgrounds – students, church extension workers, peasants and farmers – have requested advice and training. Around the project area, a reduction in land exploitation is already noticeable as farmers have begun to use the sustainable agriculture techniques pioneered by the Vacaria project.

Source: María José Guazzelli, ADFG.

TRIBAL WOMEN IN IRAN

In 1974, a pilot project on rural development was started in the mountainous region of Lorestan in the east of Iran. The local population, most of which was recently nomadic, numbers 35,000. The main objective was to promote sustainable development of natural resources through the use of appropriate technologies. It was hoped that external dependence, a dominant characteristic of underdevelopment, would gradually be eliminated.

A second objective was to help the women of that recently settled society to reclaim the important role they had played in the nomadic society of their past – without destroying the family nucleus. In contrast to their role in sedentary agricultural societies, women in nomadic communities play an active part in making decisions. Some of these societies could even be called matriarchal.

The project began by selecting extension workers, particularly women, from the local population. The workers were divided into four units: hygiene and health, agriculture, education, and rural industries and domestic economies. Health was a top priority: 16 extension workers – half the total group – were responsible for women and children's health and family planning.

The agricultural unit aimed to promote sustainable development through the use of scientifically appropriate cultivation and animal husbandry methods. The unit also tackled the reduction of malaria mosquitoes by relocating the habitat of the larvae, by improving human and animal hygiene and by introducing suitable irrigation techniques. In their nomadic past, the population had used the moulded dung of cows as domestic fuel; it was now necessary to find an alternative. The agricultural unit installed a biogas plant which met several goals: energy supply, environmental care and, most important, the recycling of organic wastes to the land.

The rural industries and domestic economies unit was formed exclusively of women. They decided to increase family living standards by promoting the weaving of handicrafts, using the traditional motifs and natural dye and methods which had been lost after settlement. The products are now a commercial success, as is the small-scale production of educational indigenous toys, which met almost all the needs of the 300 to 400 children in the district.

The Lorestan project has succeeded. Women in the health unit play an important role in family planning, disease prevention and simple healing. The education unit pioneered innovative literacy techniques and the use of a curriculum based on local culture. Women's living standards were raised as the rural industries and domestic economies group started to make and market products. Only in the agriculture unit did the presence of women extension workers create friction: men thought that it was abnormal for the four women to drive tractors, work the land, help build the biogas plant and treat animals. But, along with the measurable improvements in agriculture, the unit showed that women could take effective responsibility in agricultural affairs and animal husbandry.

Their success shows how easy it is, with a small effort, to give women confidence and to raise their awareness of their capabilities. Today the women of the region work outside the home without problems and possess a craft skill. Twelve years after the start of the project – twelve years of revolution and war – there is a noticeable difference between the women in the project area and those in neighbouring regions.

Fundamental to the success of the experiment has been its method of dialogue between project officers, extension workers and the people. At the end of each week, extension workers went home and discussed the issues with the people in their villages; they returned with many new ideas. This method requires a certain level of awareness among peasants and development workers, and sensitivity on the part of the researchers themselves. The principal goal has been to change hierarchical relationships into horizontal ones, and to formulate needs as objectives, classified in order of priority. The people were encouraged to determine their own future. Foreign extension workers were replaced by indigenous specialists, and the freedom to speak openly and to criticize was protected.

The political situation in Iran still does not encourage group discussions about solving rural problems, but the project has had a major impact. Women in the area are now openly discussing politics, their own emancipation and their political opinions with outsiders. The same women, in 1974 when the project began, would not have met foreigners nor, even if they had political opinions, would they have been able to express them.

Source: Khadijek Catherine Razavi Farvar, Rural Development Associate, Centre for Ecodevelopment Studies and Application, Tehran, Iran.

HORTICULTURE IN GOLGOTTA SETTLEMENT, ETHIOPIA

For many years, the Ethiopian countryside was decimated by a growing number of peasants applying "slash and burn" methods of cultivation. Together with constant forest fires, "slash and burn" stripped the environment of its vegetation. Adverse climatic conditions in the lowlands forced 85 per cent of the population to move up into the highlands. The country was exposed to repeated drought and famine.

In 1974, the Relief and Rehabilitation Commission (RRC) was created to coordinate efforts to repair the environmental devastation, and to initiate a programme of rehabilitation and preventive action in a number of new refugee settlements. For women, the programme was pretty much a failure. For one thing, the dependent status of most women worked against their full participation in RRC programmes – a social reality that was overlooked by the organization. For another, the structure of the settlement programmes excluded many women: only widows and women deserted by their husbands qualified for "settler status".

The Golgotta Settlement Horticultural Initiative grew out of the concern that women refugees were not taking part in the RRC development projects. The Revolutionary Ethiopian Women's Association (REWA), founded in 1981 and backed by women's groups and the FAO's Freedom from Hunger Campaign Action for Development, was given the go-ahead to start a pilot project. After assessing women's needs, REWA decided on horticulture: this would obviously provide them with food, and the fact that women would be working together would help alleviate the loneliness and insecurity of life in a settlement camp.

Golgotta, a settlement 180 kilometres from Addis Ababa, was chosen as the project site. REWA was allocated five hectares in one of the Golgotta camps with easy access to water, and promised another ten hectares if the project succeeded. RRC provided the training for both settlers and project staff.

A hundred women were involved. They began to plant immediately – watermelon, red beet and carrots – covering just over half the allotted area. In their second venture, the women planted onions, peppers and tomatoes. Harvests were good and the produce was sold locally; lack of transport precluded travel to other markets.

The women believed that the initiative was worthwhile: it gave them a chance to earn an income, and offered them, as refugees, some stability. A day-care centre, built with the help of OXFAM, UK, enabled them to work without worrying about their children. And, importantly, the pressing problem of providing food to families in an area of famine was beginning to be solved.

Source: Mutemba (ed.), 1985.

CHAPTER THREE

The Invisible Water Managers

Water is the source of all life (Koran).

In 1980, the United Nations proclaimed the period 1981 to 1990 the International Drinking Water Supply and Sanitation Decade. The UN called upon its member states and specialist agencies to "promote full participation of women in the planning, implementation and application of technology for water supply projects". The UN also urged the "organs, organizations and bodies of the UN system concerned with the Decade ... to take fully into account the needs and concerns of women" (UNICEF-INSTRAW, 1985).

Since the UN Water Conference of 1977 and the subsequent launching of the Water Decade, awareness of water as a critical issue in development has increased. Water is acknowledged as the basis of life; securing a safe and adequate supply of it is now a major task for every government. But the fact that women have specialist knowledge here, know where to collect water and how to cope when supplies are scarce, has been consistently neglected in development programmes. Several factors restrict women's influence over this area of their lives. Cultural traditions, for example, ensure that women in many societies are not permitted to intervene in decision-making, especially at the higher levels. Male heads of household decide where to build the family home, without necessarily considering the distance to water sources; water collection is not their concern. Often, ownership limits the access that women have to a water source. In Sri Lanka, for example, access to wells is determined by the caste or religious group to which a person belongs. In many societies, household budgets are controlled by men and credit facilities are not often available to women to allow them to make improvements themselves. Nor is their participation encouraged in projects which require technologies that only men have been trained to apply. Educationally, they are severely disadvantaged by high rates of illiteracy. In all, women are much less able to express their concerns or their considerable knowledge, certainly in writing. Communication with development planners, therefore, is a near impossibility.

This chapter looks at the position of women as water managers. It charts the degradation of water supplies, the traditional role women have played and the effects upon them of changes in water management. The conclusions and case studies suggest that water projects can be devised which both involve women and benefit the community as a whole.

GLOBAL DISTRIBUTION

Although 70 per cent of the earth's surface is covered by oceans and seas, fresh water is limited: it makes up less than 3 per cent of the total water mass. Most of this is ice or snow: much less is easily available for people to use. Nor is it well distributed over the world. The amount of water available in any area is limited by the hydrological cycle and by the size and development of the local population (World Resources, 1986). Whereas the total volumes of fresh water existing locally have hardly changed for centuries, water is now used very differently, and in ways which condition how much of it can be available for families. In developing countries, most water is now used for irrigation, although about 10 per cent goes to industry and an increasing amount is used for municipal purposes (UNEP, 1982). All these uses compete with family needs.

Drinking water comes from rain, rivers, streams, lakes, springs, wells and other groundwater sources. But although groundwater is potentially an important future water source for developing countries, only small amounts are presently economically exploitable (World Resources, 1986). Meanwhile, people in many parts of the world must depend upon rain as their main supplier of fresh water. But here, too, the quantity and quality is highly variable, especially in the drought-prone countries of north and sub-Saharan Africa, the Arabian peninsula, southern Iran, Pakistan and western India.

ENVIRONMENTAL DAMAGE

Droughts have always existed in certain parts of the world. But more and more evidence, particularly since the most recent Sahelian drought, suggests that these are not entirely natural (Wijkman and Timberlake, 1984). According to some climatologists, human interference with the environment may itself prolong dry periods, and many land management practices damage ecosystem balances which greatly influence the availability of water resources.

Overcultivation of cropland, overgrazed rangelands, deforestation and irrigation all change fertile agricultural land into salty, barren deserts. Stripping the vegetation exposes soils to the desiccating effects of solar radiation and the eroding impact of rain. High rainfall intensities may cause the soil to seal so that water flows off the land, causing severe erosion and downstream flooding. This erosion, the lack of water upstream and the downstream inundation can destroy productivity on a vast scale – it is estimated that deforestation and overgrazing alone are turning six million hectares of cropland into desert annually (UN Commission, 1987).

Irrigation now accounts for the largest single share of global water use – 73 per cent. But although irrigation is used extensively to increase crop yields, its efficiency is often low. In Asia, it is not uncommon for 70–80 per cent of the water drawn from a river for irrigation purposes never to reach its intended destination (World Resources, 1986). The heavy use of groundwater in dry areas depletes waterways and lowers the water table – and natural recharging takes much longer than depletion.

In northern China, the water table in some areas is dropping by four metres a year. "The depletion of groundwater resources in India, through deforestation and the other consequences of large irrigation schemes, has left 23,000 villages without drinking water" (Nalni Jayal, 1984 IUCN General Assembly). The improper design and management of irrigation projects has also brought salinization, alkalinization and waterlogging to fertile lands. Globally, salinization may have damaged 1.5 million hectares of agricultural land (World Resources, 1986).

In other words, changing land uses aimed at producing economic benefits can cause water shortages for many people, and often create associated pollution problems elsewhere. For example, in the manufacture of coffee, much water is used and polluted which people living downstream from the coffee plant are obliged to drink.

Pollution is now a major concern of water management. It derives not only from industrial processes and urbanization, but also from agriculture. In Malaysia, more than forty major rivers are so polluted (with the effluent from oil-palm and rubber industries and sewage) that they are almost devoid of fish and aquatic mammals (World Resources, 1986).

WOMEN'S WATER NEEDS

Even in the urban areas of developing countries, only 25 per cent of people have access to an in-house or courtyard water source; in some rural areas, safe drinking water is only readily available to a tenth of those who live there. The World Health Organization estimated in 1980 that more than 70 per cent of the rural populations of Kenya, Tanzania and Angola have little or no access to safe water. So limited is the access to water that some Third World women spend up to four hours every day collecting it. They do not have vehicles to carry it, as men so often do; they must transport the water on their heads.

Water is needed for many purposes in the household – sanitation and waste disposal, child care, vegetable growing and food processing. Women not only collect it for domestic purposes, but also sometimes for economic use. They may keep animals to complement the family diet and to earn money through their sale and their products. Water is also needed for crop growing and the brewing of beer, to soak seeds before they are planted and to prepare food for hired labour or for neighbours working on their plots. The provision of adequate water is an essential prerequisite if women are to become more effective income-earners. Time saved in water collection directly influences their ability to be successful in these activities, which provide them with a better chance of feeding themselves and their families and so improve their health and potential productivity. The time saved in water collection benefits not just women themselves but usually the rest of the family and indeed often the whole community. Studies have shown that poorer women, who spend more time on income-earning activities and therefore have less time for water collection, often have to accept water of a lower quality, which threatens their health and welfare.

Women as water managers

Women have to decide:

- where to collect water, how to draw, transport and store it;
- how many water sources can be used and on their quality for various purposes – for drinking, washing and in the kitchen; and
- how to purify drinking water using simple techniques (such as filtration) or materials available from the environment.

The importance of water quality for health is clear: most human diseases are transmitted by water (cholera, typhoid, infectious hepatitis) or are otherwise water-related (bilharzia, guinea worm, malaria, sleeping sickness, yellow fever) (CIDA, 1985). The evidence is that women usually take a great interest in health care. They are the ones concerned about proper waste disposal; they look after personal hygiene and take responsibility for the cleaning of latrines, the washing of clothes and dishes and the house cleaning. Because all of these tasks require water, women have established ways of reusing waste water to conserve supplies (van Wijk-Sybesma, 1985).

Over centuries, women have acquired extensive knowledge about water quality, health and sanitation. It is a knowledge they share, especially with their daughters and with each other, for women continually exchange information on these subjects at their meeting places – often the water source itself. Drawing on this knowledge, women often create their own effective primary health-care networks. (Formal health-care centres are not always accepted by local women, especially when they have not been consulted or involved in establishing them.) Through the generations, women may have developed unique customs in regard to water collection. In segregated communities, for example, women are sometimes not allowed to be seen in public. Their daughters therefore bear the burden of water collection, causing them to forsake schooling and other youthful activities.

Any effort to improve local water supplies must take account of these related issues, both general and specific. Not only do women need sources close at hand to save them valuable time in collection, but water points must continue to play their part as informal meeting places where women can exchange information and learn from each other. Moreover, educating women towards a better understanding of health care, nutrition and sanitation is likely to be most effective if the educators build on these local village networks.

Urban areas

Most women in cities depend on public faucets. The supply of water is not always continuous; waiting times can be long. Poorer city-dwellers use public water faucets to bathe, but this is difficult for women where there is no privacy. In the slums and on the outskirts of the cities, there are few or no public water points: water must be collected from sources outside the city or from vendors whose prices can be high, and whose water quality cannot be guaranteed.

Waste disposal is a major problem. In densely populated areas without latrines, many women suffer from waiting to find a suitable time and place for excretion. They often have to walk long distances to find a private site, or they must attend to their needs after dark, with all the personal safety risks that entails (van Wijk-Sybesma, 1985). But there are pioneering projects – for example, Pakistan's Baldia project discussed in Chapter 6, and the case of Rochina, Brazil on page 39.

The response of development agencies

Water projects designed to improve local conditions and to help families and small industries still take a low priority in development programmes. Generally, governments and donors favour large-scale, prestigious water schemes. But even where local projects are important, water points are often decided upon and introduced by men (UNDP, 1985). Some of these, because of the nature of their ownership or the arrangements made for their management, are not accessible to women. Handpumps, for example, are sometimes too heavy for women and girls to operate. Development agencies and their engineers are still installing schemes without giving enough attention to those who will implement them. It is often assumed that women cannot be given the responsibility for maintaining a water point. Yet, where women or women's groups are in charge, the evidence is that they maintain water sources well. Where they are trained in management and repair, water points often function more effectively than when the responsibility depends upon a technician living some distance from a local community.

International Drinking Water Supply and Sanitation Decade

Two years into the Decade, it became clear that women were still not adequately involved in its activities. In 1982, a Steering Committee for Cooperative Action established an Inter-Agency Task Force on Women and Water to develop a strategy for evaluating and emphasizing women's participation in the work of the Decade. It was concluded that their participation would increase only if better communication took place with existing women's organizations, if technologies were chosen more carefully, if more women were trained in the maintenance of equipment and if water and sanitation projects were seen as part of a comprehensive local programme of health and welfare improvement.

Education and training are being stressed. As well as learning about the maintenance and repair of water points and sewage systems, family members, especially children, have to be shown the importance of properly disposing of their faeces and washing their hands (particularly before they deal with food). As part of the Decade, studies (using local interviewers) are being made of existing patterns of water use so that these can be made more sanitary and efficient (UNICEF/INSTRAW,

1985). Special meetings of women will be organized to find out their problems, their knowledge and their wishes. Local women as well as women teachers, nurses and midwives are being trained to pass on the knowledge, and advice handbooks will be prepared for distribution among women's groups. Special attention is being paid to the effective training of women in water management – ensuring child-care arrangements so that women can attend training sessions, involving them in the development of their own training materials, and in the coordination of training efforts among field staff concerned with water, health and welfare. Throughout, the emphasis will be on strengthening existing community structures.

More appropriate technologies, based upon local knowledge and local materials, are likely to be introduced as the Decade proceeds. Already there is closer involvement of village communities in the planning and implementation of new water sources.

In small ways, the Decade is already showing results. By 1990, it is estimated that almost every village in Thailand will be able to provide a minimum of two litres of safe drinking water each day for each person, as well as basic sanitation for every household. This success is due to the local, decentralized approach of working through district and village organizations with funds allocated directly to them and to appropriate supporting institutions. In a number of countries, UNDP is supervising a regional programme of water improvements involving women in the projects' design and execution. The programme will include research on needs and the education and training of local workers to be sensitive to women's requirements (UNDP, 1985). UNDP intends this to be a demonstration programme of the value of community participation in water management.

CASE STUDIES

THE WOMEN'S DAM, BURKINA FASO

For over a decade, aid agencies spoke of the great potential of the Yatenga plateau, Burkina Faso, for irrigated agriculture, cotton growing and grain production. The problem, year after year, was severe water shortages. Rains were erratic; when they came they would quickly disappear deep into the earth. The groundwater table was such that the only practical means to a permanent water source would have been drills and pumps. But in the villages, many traditional means of catching surface water were employed. These consisted of hand-dug drinking holes, wells and small earthen dams. The wells and dams were built mainly to meet the daily needs of people and their animals, for water shortages brought great hardships.

The villagers of Saye talked for years of building earthen dams to catch the rainy-season waters and hold them longer into the next dry season. But still in 1979 nothing had been done. Finally, during one of the most severe water shortages known, village women organized a meeting. Three representatives approached village men to state that if they could not be persuaded to help build a dam, then the women would build one themselves. If this could not be done, they were resolved to return to their parents' villages where there was more water. The women were not willing to continue carrying water over long distances, pounding grains and gathering fuel. The men realized the seriousness of the situation. And after all, the elders had placed their faith in the ancestors and Muslim God to bring back the rains, but none had come. Perhaps the women were right; it was time to try something new.

The date was set for work to begin. Minata, leader of the women's group in Somiaga, a neighbouring village, came to help Kadisso, traditional healer and leader in Saye. Youth groups arrived with long drums strapped to the backs of their bicycles. The older *griots* (traditional court singers) carried drums in their arms as they walked. Villagers came in donkey carts with loads of women and children. Eventually, hundreds of people from more than three villages had gathered. It was the dry season and there was little other work to be done. Each had his or her task. The old men who could not work sat under trees, watching toddlers and encouraging others. Some grandmothers worked in the gravel pits, loading baskets and pails for the younger women to carry four kilometres to the work site. The final day of dam work was celebrated by feasting and dancing. Everyone was invited and everyone came. Among the honoured guests, Minata from Somiaga received the greatest attention. Crowds recognized the familiar tribal scars on her face as she stepped up to sing with the women of Saye. They sang of beautiful trees and plants which would grow around the dam. Minata composed a song:

We worked together to gather stones
We made a dam;
All the men who travelled to Mecca say they gathered stones
to throw at the evil tombs of disbelievers
Like them, we gathered stones
But we were going to build a dam,
A future for our children, our village, for Burkina Faso and all of Africa.

When she stopped, the women encircled her and would not let her leave. Festivities continued long into the afternoon until clouds gathered overhead.

Then the villagers dispersed and, a few hours later, the first rains of 1981 fell in Saye.

In the nearby village of Somiaga, villagers consider Saye to have been a testing ground for their own dam. One man said, "We learned from the women in Saye that dams should be built; we also learned from their mistakes." But the mistakes did not discourage them from trying new things. With 1,600 inhabitants, Somiaga had almost four times as many workers as Saye, so their dam could be finished in less than three months. Others would also come to help.

The importance of self-reliance is one of the most significant lessons of the dam-building projects. The integration of water and sanitation activities with other dimensions of development is leading to integrated rural development. The dams serve as a point of entry into the broader aim of reviving the spiritual, social and cultural vigour of traditional societies.

Source: Soon Young Yoon, 1983. Adaptation by Stephanie Flanders, IUCN, 1986.

WATER FOR HEALTH IN KENYA

KWAHO – the Kenya Water for Health Organization – is a consortium of NGOs established to respond to women's self-help efforts for water and sanitation improvements. KWAHO, with other groups, is developing the methodology and materials for community liaison and training. The programme is supported by the Kenyan government, the United Nations Development Programme (UNDP), the World Bank and the United Nations Development Fund for Women (UNIFEM).

One example of its work is in the Kwale District, south of Mombasa, where KWAHO is organizing a cooperative programme to train women to use and maintain simple water systems. In this pilot project, villagers are taking responsibility for all phases of handpump installation and maintenance. Cooperation between project staff and the community has been developed by five female extension workers and a process of village-level decision-making. Discussions have been held about villagers' needs for water, including specific items such as the number of wells required, their siting, and the ownership of wells and handpumps. Wells have been constructed of local materials and with labour provided by the villagers, from whom funds have been collected for maintenance. Training has been provided in handpump maintenance, group organization and book-keeping. A complementary project is training women in health, water use and maintenance.

Because of the success of the pilot project the methodology is being applied throughout the entire district. Extension workers have been selected to spread messages about health and water use to local communities and 24 female trainees from the Kwale region are participating in the general training programme. Village-level training courses are scheduled, and household surveys to collect data on water use, storage and sanitation facilities are in progress.

The project is being carefully evaluated and, from time to time, workshops are organized to assess the effectiveness of the entire programme and to decide how it can be expanded to other parts of Kenya and possibly other developing countries.

The Kabondo Women's Group

KWAHO also supports the Kabondo Women's Group in Wangapala, a small town in the densely populated area of South Nyanza, western Kenya. Here, there are some five hundred semi-permanent, tin-roofed houses and an average family has twenty members. Most of the people are farmers and everyone owns a piece of land. The area has a good rainfall so crops of coffee, groundnuts and maize are sown twice a year.

During the Mexico Women's Conference in 1975, it was decided to mark the beginning of the Women's Decade with a "Women and Water for Health Project" in western Kenya. The Kabondo Community Self-Help Group was already active in the area, helping small farmers to cultivate in time for the rains. Most members of the group were women; occasionally some husbands joined. The community faced a serious problem of access to drinking water. The nearest source, a spring, was at a distance of six miles. Every day, women had to trek down the hill and then climb back with the load of water. It was a time-consuming and back-breaking job.

Through the Kenyan office of UNICEF, women from all over the world provided the funds and technical assistance for the project. The villagers voluntarily dug the trenches and a pipeline was laid. UNICEF supplied the equipment needed to pump water from the spring, which was at a lower level than the village.

There are now four water points in the village and the government and local authorities assist with the pump's maintenance. Usually, the headmaster of the school or the chief of the area takes on the responsibility for the daily running of the pump. There are no paid workers: all the tasks are undertaken by the community and the members of the Kabondo group. Water is available throughout the day at all points. Every family contributes about thirty shillings a month towards the costs of diesel fuel and maintenance. To improve the facilities available for health care, there is a plan to extend the pipeline to the Health Centre, which is three miles from the school.

The project received financial assistance from national NGOs such as KWAHO and Zonta Club, and initial funding from UNICEF. The project today is completely self-supporting. Women used to spend most of their day fetching water and collecting firewood. Now, they have time to engage in other activities. KWAHO provided the women with tree seeds and they have started to grow fruit and firewood trees around their homes. Some have taken to vegetable cultivation, selling surplus produce to increase family income. They are also involved in other income-generating activities, such as bee-keeping and the milling of maize grain to make posho.

People are healthier because of the addition of fruit and vegetables to their diets. Because of the abundance of water, the environment and the children are cleaner. The community spirit among members has increased and they are engaged in many group activities. They have started a land donation scheme whereby each member, by rotation, donates a piece of land for the group's use for a limited period. All members work together and grow cash-crops such as groundnuts and bananas and the money earned is banked and used for starting new projects like the installation of the posho mill.

This project shows how fruitful it can be to work with an existing group when such a group considers a project to be its responsibility, not an imposition from some international agency. It is essential that proper leaders are identified: the ultimate success and speed of implementation depend upon the quality of the leadership. Support is also important, including that from government officials, chiefs, sub-chiefs, social workers, teachers and – of course – the people.

Sources: Prahba Bhardwaj and KWAHO, 1984; UNDP, 1985.

WATER FOR ROCHINA, BRAZIL
Rochina is one of the shanty towns of Rio de Janeiro. Its population lives in poverty. Many in the illegal settlement live in wooden shacks, though others live in brick and stucco houses with a water supply, paved streets and electricity. But throughout the shanty town, raw sewage and garbage accumulate next to people's homes, along paths and on vacant land. The filth is washed down into open drains in the rainy season and often these flood the low-lying parts of the town, contaminating water supplies. Children usually fetch and carry the water for their families and take away the refuse, often playing in it.

Community groups have always been active here in cleaning up the town and improving its sanitation. But alone, they were never very successful. In 1979, the local authority together with UNICEF began an urban development plan for Rio, with the emphasis on improving living conditions in the shanties. Community participation – involving women and their groups – was at the heart of the strategy. A successful method of collaboration evolved between local community groups and the government.

At first, although the shanty residents agreed that health and sanitation improvements were needed, they were wary of outsiders. In 1981, three pilot projects began: on basic sanitation, community schools and health care. To build up local trust, part-time community workers were recruited to identify the priorities. It was decided to add to the sewage system that local people had already started to build; with government and UNICEF help, the new system was completed, serving 120 families. The area around the public water tap was drained and paved and the municipality began a garbage collection system. In all this, the local community provided labour, equipment and materials and organized the work, keeping the cost of the project low.

Now, the community is no longer suspicious of outside agents (and is demanding more technical support). The community workers proved to be very effective helpers and good motivators of local residents. The project demonstrates the value of combining the efforts of community, government and UNICEF. UNICEF expects gradually to withdraw from the project, shifting the responsibility to the local authorities. Wide publicity has encouraged other shanty towns to become active in seeking improvements.

Source: Chauhan *et al.*, 1983.

CANAL HURTS COLOMBIAN WOMEN

In fishing villages along the Pacific coast of Colombia the gathering of cockles (locally called "piangua") is one of the major sources of women's income. Their livelihood is threatened by the Esteros Project, which involves the construction of some four thousand kilometres of canals to link the tidal creek system between Buenaventura and Tumaco. Preparations are already under way for dredging and excavation, although this will be necessary only in shallow stretches that, it is estimated, occupy less than 10 per cent of the total canal alignment.

As part of an environmental impact study (which is legally required in Colombia) by the National Colombian Institute for Environmental Management (INDERENA), a special survey was made of the environmental impact of dredging and excavation on the mangroves and their associated cockles, and the piangua fishery. Socio-economic data were obtained through interviewing the local population, village leaders and women's groups; several case studies were carried out.

Although it was not possible to quantify the effects of the canal alignment on other natural resources, the research team concluded that the destruction of the Rhizophora (mangrove) forest would have an effect on the productivity of estuarine organisms that depend on that habitat during part of their life cycle. More severe, however, would be the impact on the income of the women in the fishing village of Salahonda. The researchers found that the piangua catch declined significantly after dredging and excavation had begun. Before the works, catches varied between 300 and 500 cockles a day, while the study indicated daily catches of 150 to 350. In interviews, the women confirmed that there had been a decline in catches. Completion of the canal works and the consequent loss of suitable piangua habitat would result in a net loss of 4.1 million Colombian pesos a year of potential income for local people.

How to limit the damage done by the canal? It was decided that only limited areas would be dredged and excavated. At the same time, local communities would benefit from the canal works because of the better accessibility to markets and lower transport costs. Finally, women's groups were set up in the project area and provided with better canoes and small outboard engines to enable the women to reach less accessible piangua areas.

Source: Netherlands Directorate General for Development Co-operation.

CHAPTER FOUR

Women and Forests: Fuel, Food and Fodder

A CORNUCOPIA OF BENEFITS

A forest is not just a stand of trees. Forests are essential to sustain world ecology and human life. They protect watersheds and regulate water flows, the absorption of rains and evaporation. They maintain the ecological balance for a regular and clean water supply and help to protect agricultural lands, especially those downstream. Through their root systems and foliage, forests play an essential role in soil protection. In many countries, forests form the basis of civilization. India, for example, became known as Arany Samskriti, or "a forest culture", and forest-based settlements produced the best scientific research and cultural writings (Shiva *et al.*, 1985). The Chipko movement in India expresses the value of forests in songs:

> What forests bring us,
> soil, water and clean air;
> soil, water and clean air,
> the basis of our life.

For about 200 million people, forests are their only home. They provide fuel and construction wood, animal products, vegetables, fruit, nuts, honey and spices, organic fertilizers, fodder for animals, medicines, and many raw materials for industry such as oils, resins, gums, rubbers, waxes, fibres, rattan and bamboo. Wood and the so-called "minor forest products" are crucial for people living in and around these ecosystems.

Forests and woodlots also provide 66 per cent of the net primary productivity of terrestrial ecosystems; tropical forests in particular account for three-quarters of that productivity with their exceptional diversity of plant and animal species. And trees outside the major closed forests – in shrubland, windbreaks and woodlots around farms – are also important, particularly in densely populated areas. In Rwanda, the woodlots associated with cultivated land and pastures – approximately 20,000 hectares – exceed the area of remaining natural forests, as well as state and communal plantations (World Resources, 1986).

Serving women's needs

For centuries women have gathered forest products. This remains an important activity for tribal societies and for many other rural households in Africa, Asia and Latin America. "The time spent in forests, gathering wood, has taught women the many uses of trees, including providing fibres for cloth, mat-making and basketry. Many trees are used as a source of food, offering vegetables, nuts, fruits and even vines. Women also know the medical uses of various trees ..." (Aloo, Nairobi NGO Forum, 1985).

In one area of Sierra Leone, women listed 31 products they gather from bushes and trees near their village (Hoskins, 1979). In traditional hunter-gatherer societies, like that of the !Kung Bushmen of the Kalahari Desert, women are major food providers through gathering in the woodlands (Draper, 1975). Anthropologist Sally Slocum (Reiter, 1975) states that the concept of "man the hunter" as the only provider of family food is quite inadequate. Of equal or even greater importance for family well-being is the role of "woman the gatherer". In the Kalahari, it is women with young children or adolescent unmarried girls who usually gather bush food. Small groups of women forage at distances of eight to ten kilometres from home. "It seems likely that !Kung men and women have similar knowledge of the larger hunting and gathering territory within their kin and affines range" (Draper, 1975). In sedentary !Kung villages, however, Draper observed that the sexual egalitarianism of the bush setting is being undermined. There, women's work is seen as "unworthy" of men.

For women, trees and forests are multifunctional, whereas men tend to concentrate on their commercial potential for timber and other goods. Trees offer "fuel, food and fodder" – the Three F's, as women say.

VANISHING FORESTS

Originally the earth must have been covered with some 6,000 million hectares of forest and woodlands. In 1954, only 4,000 million hectares remained: a loss of 30 per cent. Tropical forests – including closed

Figure 3:   *Summary of Women's Interests in Forest Resources*

Primary Tree Products

Daily fuelwood collection near the household. Concern over availability of preferred species. Interest in access to building poles for local use.

Secondary Tree Products

Major involvement in collecting human food and having available fodder for small animals near home site. In certain areas where cattle are kept at the household, women are in charge of gathering fodder.

Tertiary Tree Products

Collect numerous products needed in the household and for barter or sale. Women's employment or extra cash income may depend on access to tertiary products as raw materials.

Soil

Use limited to areas near household. Special interest in soil quality in gardens and in fields with subsistence crops.

Source: Rural Women, Forest Outputs and Forestry Projects. Discussion Draft. FAO, 1983.

forests, open forests, shrubland and forest fallow – now cover under 3,000 million hectares. Latin America still has most of the world's tropical closed forests; Africa carries two-thirds of all open tropical forests. The percentage of wooded land is lowest in Asia. But there is considerable controversy over available estimates of the rate of tropical deforestation, and the remaining area of forest. The FAO projected in 1980 that 150 million hectares (or 12 per cent) of the remaining tropical forests and roughly 76 million hectares of open tropical woodlands would be gone by the year 2000 (World Resources, 1986).

Deforestation – the clearing of open and closed forests – is one of the world's most pressing land use problems. Selective logging of closed forests causes enormous degradation. Conversion to agricultural land is also a major cause of deforestation. Other important factors are commercial timber logging, planned migration and resettlement, land speculation, large-scale construction projects, and the expansion of commercial ranches. At a local level, forest fires, growing demands for firewood and fodder, and grazing contribute to the loss of forest. The consequences are severe: disturbance of water systems, for example, results in flooding and drought and, as in the Himalayan and Andean regions, destroys fields and women's livelihoods.

In **Africa** 1.3 million hectares of closed broad-leaved forests have been cleared annually during the past ten years, and 2.3 million hectares of open woodland are being lost each year. More than half of the loss takes place in West African countries such as Ghana, Guinea, Ivory Coast, Liberia and Nigeria. Madagascar alone accounts for the greatest East African loss of forest: 200,000 hectares each year from tree-felling and bush-fires (Ramanankasina, 1985).

From 1976 to 1980, 1.8 million hectares of closed forests were degraded in **Asia**, especially in Burma, India, Indonesia, Laos, Malaysia and the Philippines. But the highest deforestation rates are those of Nepal (3.9 per cent per year) and Thailand (2.4 per cent per year). Over the past thirty years, Himalayan watershed forests have declined by 40 per cent.

Each year in **Latin America**, 4 million hectares of forests are cleared or converted to other uses. Brazil alone accounts for 35 per cent of the loss, but in almost all other countries deforestation is high and rising, particularly in Costa Rica, El Salvador and Paraguay. By 1978, only 10 per cent of Haiti remained under forest, and the decline has since continued (World Resources, 1986).

In tropical regions, deforestation rates have outstripped reforestation by up to twenty times in recent years. And "industrial plantations, which were initiated during the colonial era, continue today in an irrational manner, having catastrophic impacts on tropical rain forests" (Ramanankasina, 1985). The present rate of reforestation is assessed to be less than 10 per cent of that needed to supply minimum needs of many countries in the Third World by the year 2000.

Are women to blame?

With so much pressure on the land, women have little choice but to use forests (Bamba, 1985). In the Ivory Coast, for example, Krou women living in the neighbourhood of the Tai Forest are driven by agricultural developments (such as cash-cropping of oil palm and coconut) and by increasing immigration from abroad to move into the forests to use the clearings left by loggers.

The same is true of fuelwood collection all over the world. As there are no alternatives, women must gather wood and other biomass for fuel wherever they can. They rarely collect entire trees; they take only twigs, smaller branches and often dead wood, so destruction is limited. The next chapter deals in more detail with energy issues.

Deforestation hurts women

Women in the Uttarrkhand hills (India) say: "When we were young, we used to go to the forest early in the morning without eating anything. There we would eat plenty of berries and wild fruit ... drink the cold sweet (water) of the Banj (oak) roots.... In a short while we would gather all the fodder and firewood needed, rest under the shade of some huge tree and then go home. Now, with the going of the trees, everything else has gone too!" (Sunderlal Bahuguna, 1984:112).

The loss of forest is accelerating worldwide, but the figures tell us nothing of the daily pressures which deforestation and the degradation of forest lands place on women.

The next chapter on energy shows how the scarcity of firewood means more hours spent collecting fuel. It also leads to changes in cooking habits, which can damage family health. The case of the Gadkharkh village in India shows how deforestation creates an unbearable situation for women (page 57). They must walk long distances, perhaps 10 to 12 kilometres, and leave before dawn just to collect firewood with which to cook. And "as the environment degrades, both fodder and water become difficult to obtain, forcing women to spend more time foraging for these basic necessities in addition to other work" (Sheth, 1985).

Inappropriate afforestation – like the eucalyptus plantations of India

– can also influence women's work. These trees absorb large quantities of surface water, so that women, the water collectors, are faced with lowering groundwater levels and a drying-up of their water sources (Shiva *et al.*, 1985).

CHANGING FOREST MANAGEMENT

With commercialization, modernization and increasing demand for wood, many of the traditional customs of forest protection have broken down. Only recently have the needs of local forest users been considered. Working with local people to control forest exploitation is necessary for regeneration and a sustained yield. Methods must be developed, argues Cecelski (1985), to balance local people's needs with national claims in an ecologically sound way.

Throughout the Third World, forestry is dominated by industrial plantations and reserve management. "The perception of forest ecosystems as having multiple functions for satisfying diverse and vital human needs for air, water, and food has been replaced by uni-dimensional 'scientific forestry'" (Shiva *et al.*, 1985). This "scientific" perspective often becomes no more than a calculation of timber yields to serve commercial and industrial demands. Ecologically, as well as socially, there are serious consequences. Monocultures replace multi-species forests, and this disturbs natural balances so that forests are no longer able to meet people's, and especially women's, needs. Massive afforestation programmes in Peru and India, for example, met industrial demands, but removed grazing and agricultural lands from local use (Cecelski, 1985).

Although natural forests are still being converted into monocultures of such species as eucalyptus – destroying the water balance, soils, ecological diversity, and the capacity to produce fodder and organic fertilizer – much attention is now being paid to "forestry for local community development" or "social forestry". The recent emphasis on basic human needs has promoted concern for the local community. Forest development, argues Williams (1985-1), should not be limited simply to managing trees or making a profit; rather, it should be devoted to managing socially valued resources to serve human needs. Thus, "foresters need to adopt a truly social vision of development".

Social forestry
The FAO defines a community approach to forestry as "any situation

which intimately involves local people in forestry activities" (FAO, 1978). This approach encompasses a broad spectrum of the features of rural life and is, therefore, far more difficult and time-consuming than traditional forestry. Local residents must be fully informed and encouraged to participate at all stages of a project.

Like commercial forestry, social forestry has seen many failures over the past years. It has proved to be effective in increasing wood production outside forests, on farms and private lands. But in many countries it has failed to benefit the rural poor, especially the landless. Social forestry in India, for example, has provided an incentive for farmers to exchange irrigated food-producing land for eucalyptus plantations. These provide neither fodder nor fuel for local communities but will be used to supply industrial textile and paper mills (Cecelski, 1985). In Sudan, planting of the Acacia Senegal has been encouraged by the Forest Department to combat desertification, but this is profitable only for a few large landowners and reduces the ability of small landowners (including women) to fallow their lands by renting fields from larger farms. In Niger, people removed trees planted under a World Bank project: they were not involved and the trees had been planted on their traditional grazing lands (Spears, 1978). Communal village woodlots were introduced in Tanzania under fairly strong political pressure at the end of the 1970s. But most of these have never been harvested (McCall Skutch, 1986).

It is clear that if the poor are to benefit from social forestry projects, they must be involved and they must share in the results of the projects. Home gardens, for example, can provide food, fodder, fuel and income. Where they are close to urban areas or industrial markets, woodlots and orchards can produce fuelwood, charcoal, poles, fruit or other items for sale to the benefit of the local economy (Cecelski, 1985).

WOMEN PROTEST

Experiences from all over the world show that women, despite their long and arduous working schedule, have a great interest in defending and restoring the forest ecosystem. India's Chipko Andolan movement is a famous example of women protesting against forest destruction.

In the Reni forests of the Chamoli district, Uttar Pradesh, in 1974, women were confronted with the prospect of 2,500 trees being destroyed by commercial enterprise. The women were alone, for their men had left home. When the contractors arrived, they went into the forest, joined hands and encircled the trees ("chipko" means "to hug"). The women told the cutters that before cutting the trees, they would first have to cut off their heads. The contractors withdrew and the forest was saved.

In the state of Rajasthan in India, clear cutting and forest pillaging were common practices for years. By 1960, the destruction of Himalayan forests had become the major cause of ecological instability in the region, bringing loss of forage and fuel and repeated catastrophic flooding. Resistance to increased destruction of the forests by state and private agencies was strong: women fasted, guarded the forests, and wrapped themselves around the trees to be felled. As a result of women's actions to preserve their environment, Indira Gandhi issued a fifteen-year ban on commercial felling in the forests of Uttar Pradesh (Shiva *et al.*, 1985).

The Chipko Andolan movement grew out of this protest. It is a movement in which both village women and men participate as leaders. It now spans the whole Himalayan region. On their marches (covering as much as 5,000 kilometres through the Himalayan mountains in India, Nepal and Bhutan) Chipko activists came into contact with the societies of remote hill areas. Their message of anguish about the ecological situation of the region began to spread as more and more people and villages became involved (Ummaya *et al.*, 1983). In Uttorkashi, for example, hundreds of women formed a procession to demand the preservation of natural mixed forests. In the Jakhur Valley, villagers appointed their own forest guards, having framed rules and working plans for the preservation of the forests in the neighbourhood.

In Khirakot, a small village in the Almara district of Uttar Pradesh, women collect fuel and fodder from the surrounding "panchayat" forests owned by the "village community". They preserved these forests very carefully – until a Kanpur contractor obtained a lease for soapstone mining in the hills. The women realized that their forest access was being hindered by mining activities, and that the forests would be killed by mine debris. And although local men were employed by the mine, women protested. "Either the mine will remain or we," they stated. Even under serious threats by the contractors, the women brought them to court. The mines were officially closed (State of India's Environment, 1985).

Women in Duagara Paiteli protested strongly when they learned in 1978 that their community forest had been sold by the male-dominated panchayat to become a potato farm. Village men believed they would become employed on the farm and that many improvements (roads, for example) would result from the project. But the issue turned wives against husbands and mothers against sons. The women refused to walk the extra five kilometres each day to fetch fuel and fodder. In spite of strong opposition and threats from their menfolk and the district administration, the women eventually saved the forest (Sheth, 1985).

Protests continue elsewhere. In Brazil, the Ação Democrática Feminina Gaúcha (ADFG) has developed into a strong environmental defender, opposing activities which destroy the Amazon rain forest. In the Philippines, the Lingkod Tao-Kalikason (Secretariat for an Ecologically Sound Philippines), headed by two women and including many women members, promotes environmental action "with a view to reversing the tragic destruction of the earth", including the forests.

WOMEN ARE GREENING THE ENVIRONMENT

> Come, arise, my brothers and sisters,
> Save this mountain ...
> Come plant new trees, new forests,
> Decorate the earth (Chipko Andolan song).

Protests against deforestation are not the only responses. Women are also leading attempts to reverse the destruction by planting trees. India's Chipko Andolan movement protests against forest destruction and acts to rehabilitate the environment. Dasohli Gram Swaraj Mandal (DGSM), the Gopshwar-based organization at the forefront of the Chipko movement, organizes ecodevelopment camps in which rehabilitation of the ecological balance by tree planting is promoted. It was after a landslide near Pakhi – which brought the environmental crisis to their doorsteps – that local people started taking an interest in afforestation work. Survival rates of the trees began to crawl upwards from a dismal 10 per cent as local women began to care for them. Now the average survival rate of Chipko plantations is 80 to 90 per cent. From Pakhi the reforestation work expanded to other villages. In 1982 and 1983 the DGSM organized 20 ecodevelopment camps. And while the camps are open to both men and women, their different interests in forestry have become evident: men often withdraw from the camps, while women's involvement is increasing. When the first village assembly was asked to choose what kind of trees to plant, men immediately replied "fruit trees". The women argued against this: "The men," they said, "would take the fruits and sell them by the roadside.

The cash will only go to buy liquor and tobacco. We women prefer fuel and fodder trees" (Agarwal and Anand, 1982).

Kenya's Green Belt Movement, started in 1977 by the National Council of Women, is a well-known example of positive development. The main objectives are awareness-raising and the prevention of deforestation by tree planting. Under the leadership of Professor Wangari Maathai, the Green Belt Movement focuses on campaigns and the local participation of women in the establishment of "Green Belt communities" and small tree nurseries. By 1982 there were 50 nurseries producing 2,000 to 10,000 seedlings per year, and 239 "green belts". An additional objective is to promote cooperation with official authorities. Hundreds of groups with memberships of between 20 and 100 women are active in this area now. Not only does the environment benefit, but women gain more power and earn incomes from the sale of seedlings. Tree-planting activities are effective awareness-raisers for the environment, particularly as they often take place during commemorations and special celebrations (Obel, 1985; see also the case study in Chapter 10).

Tree planting is a focus of many other Kenyan women's organizations as well. Women's groups are actively involved in the Kenya Energy Non-Governmental Organization (KENGO), a platform of more than 200 NGOs dealing with energy issues. Although their main objective is to safeguard energy supplies through, for example, information about fuel-saving stoves (see Chapter 5), another programme deals with reforestation using indigenous trees. After collecting information from local women on the medical, cultural, ecological and economic value of trees, KENGO passes it on to women's groups and others by means of district workshops, exhibitions, radio programmes, newspaper articles and publications. It also sponsors a tree seed project, in which people are trained in collecting, handling and pre-treating indigenous seeds for planting. In all these ways, KENGO hopes to promote individual and communal reforestation with indigenous trees (Musumba, 1985).

Maendeleo Ya Wanawake, another important umbrella organization in Kenya, encompasses 80,000 women's groups and more than 300,000 members all over the country. Maendeleo's Environment Conservation Programme has tree-planting projects in more than nine Kenyan districts, including Nairobi. In Bolivia the Young Women's Christian Association (YWCA) has started a project called "Preservation and Care of the Environment". One of their activities will be the planting of 7,000 trees around the city oí Potosí (Y's EYEs, 1986:1). In South Korea, the Mothers' Club of Korea raises funds by growing and selling tree seedlings. And in Lesotho, women's groups plant woodlots for their own use (FAO, 1983). Similar initiatives, as described in the case studies, are under way in the village of Gadkharkh, India, and in Nepal.

Although women's efforts in reafforestation will not reverse the trend of diminishing resources, they can help to address deeper injustices and inequalities. Lori-Ann Thrupp, formerly associated with the Institute of Development Studies, Sussex, England, maintains that "If oriented in a proper way these [women's efforts] can lead in successful directions" (Thrupp, 1984).

INVOLVING WOMEN IN FORESTRY

Paula Williams of USAID, one of a mere nine per cent of female participants attending the IX World Forestry Congress in Mexico in July 1985, argues that despite women's widespread use and management of forest resources, forestry and forest development policies have largely ignored women and their activities. Only their role as firewood consumers has received much attention. "Understanding the interactions between women and forests is vital for a comprehensive view of forest resource use," she concludes, "and women should be involved in all aspects ... of forestry" (Williams, 1985-2).

At the World Consultation on Forestry Education and Training, it was noted that there are some forest activities, such as nursery and plantation work, in which women are engaged, but they continue to be excluded from forest management and use operations, such as surveying, logging, protection and control. Discrimination against women workers in terms of pay is common (Foster, 1986). Even in the "new" approach of social forestry, women are largely excluded. When approaching local residents, foresters and extension workers often speak only to community leaders and village councils, most of whom are men (Hoskins, 1979). Globally, the number of professional women foresters is still very low, although they are beginning to emerge from training in several Third World countries, of which Burkina Faso is one (Williams, 1985-2). While increasing their numbers does not automatically mean that women's needs and priorities are adequately taken into account, professional women can serve as catalysts for change.

Kenya's first woman Forest Officer, Theresa Aloo, was appointed in 1971, originally as an Assistant Conservator of Forests. In 1986, the total number of women holding such positions was nine, namely four assistants and five foresters. A few are found in educational and training institutions, but as Theresa Aloo explains, the traditional approach to forestry has made it difficult for women to join the profession (see page 63).

In India, three women have recently been admitted to the Dehra Dun Institute of Forestry. But while the training of professional women is important, their future job opportunities must also be considered. After the term of study, it is often difficult for a woman to find employment in forestry. The job of forest extension worker also seems to be virtually unobtainable. The Indian College for Home Sciences, which has a majority of rural girls on its roll, and the women's college in Jabalpur both maintain that their students cannot find appropriate employment. Yet these local (often tribal) girls, who understand local problems, would communicate far more effectively with village women than the professional men who usually hold these positions (Bhatty, 1984).

Women's participation in forestry and forestry-related activities in peninsular Malaysia is similarly unsatisfactory (Women in Forestry, 1984). In that country, nine female Sub-assistant Conservators of Forests were employed by the Federal and State Forestry Departments in 1984. The total number of departmental employees is 4,500. But it is estimated that 60 to 70 per cent of the production line machine operators in private veneer and plywood mills are female workers. The majority are young women; most of them receive lower pay than male workers.

No detailed studies of women's participation in forestry and related activities have been made in the Philippines, but there are now women foresters and forest rangers. The University of the Philippines at Los Banos College of Forestry had, in 1984, 148 women graduates in its forest ranger curriculum and 89 in its bachelor's degree programme for forestry (Cruz, 1984).

In Thailand, as recently as 1975, women were still not allowed to enrol in a school or college of forestry, as the subject was regarded as too strenuous for women. But between 1975 and 1984, 48 women had been admitted to the Faculty of Forestry, Kasetsart University. The total percentage of women workers in forestry is some 20 per cent (Women in Forestry, 1984).

Improving social forestry
It might be thought that "social forestry" would show a different picture of women's involvement. Yet the FAO reports that "Although

women in rural areas are directly dependent on forestry-related resources, many forestry projects are designed without mention of women and without any recognition of the impact the proposed activity will have on them" (FAO, 1983).

Women's participation has been critical to the success of many community forestry projects, but few such social forestry programmes include women in sufficient numbers. An evaluation of cookstoves and social forestry programmes in Gujarat, India, showed that the wood produced under these programmes is being taken by dealers to the cities, where it is used for construction or sold to townspeople as fuel. Nothing is left to relieve the scarcity of wood in the villages. Yet the Gujarat programme is considered to be a pioneer case of social forestry because of its high production levels. "The prestige and fame of the sponsors ... seem to count more than benefits to local residents" (McCall Skutch, 1986).

In Senegal, the timber in village woodlots reserved for meeting household fuel needs (and reducing women's workloads) was planned to be sold as polewood by men in village councils (Cecelski, 1985). Project managers report that including women is not always easy, but since women are so essential, one needs to examine very carefully how their participation can be guaranteed. Its importance is evident if one considers their knowledge of local situations. In a seminar on women and development held in Burkina Faso, for example, women unexpectedly referred to the environmental deterioration in their country and to all the consequences of eucalyptus plantations. They demonstrated a profound expertise, concern and awareness of the local situation (Hoskins, 1979).

Women's work on forestry projects ranges from paid employment on food-for-work programmes to voluntary activities. In Burkina Faso, for example, women's contribution is almost always unpaid. However, participation should not mean just cheap labour. Voluntary labour is only beneficial "when it contributes to the participants' own development; that is, increasing the control that individuals have over their own lives.... Development thus constitutes an enhancement of people's ability to use resources in problem-solving strategies" (Williams, 1985-1). Before starting any project or programme, a clear analysis of women's needs and situation is necessary, for when women are already overworked, additional voluntary effort should not be expected from them except when it will ultimately lighten their burden. Women in Guinea, for example, chose to plant nurseries and woodlots on a communal basis, but as they were already overworked, they asked for men's support – and received it (FAO, 1983).

The Kenya Woodfuel Development Programme (KWDP) has systematically analysed fuelwood supplies in Kenya and, on the basis of its conclusions, developed strategies to alleviate Kenya's fuelwood shortage. For the Kakamega district, a self-sustaining system of tree planting was developed to contribute to fuelwood supplies. Using the findings of surveys of agroforestry activities and the cultural background of district inhabitants, the KWDP has designed an approach that accommodates indigenous expertise with traditional beliefs and taboos rather than attempting to impose potentially unacceptable solutions on the people. In this process, women especially are encouraged to formulate solutions, so that these will not conflict with traditional values. Technical help from trained local extension staff is available whenever it is needed. The project is linked to an awareness-raising programme, using popular mass-media techniques such as drama, pamphlets and posters (Chavangi *et al.*, 1985).

Finding out women's needs
Marilyn Hoskins (1979) defines the major issues to be examined when involving women in local forestry projects:

- What are women's specific problems in gaining or retaining access to land or using tree products?
- What specific time, financial and other factors constrain women's participation?
- What measures ensure that women benefit from projects?
- What different social structures allow women to participate as individuals or in groups?

It is not enough to generalize, she concludes; each of these questions must be answered with information from the local community and specific programmes must be designed with local residents and institutions.

Women's participation is constrained in many ways: competition for land; problems over land tenure; lack of time left after domestic duties; cultural taboos and lack of familiarity with forestry. Moreover, women do not have institutional and organizational support. Legal restrictions, an absence of legal entitlement to land, and the low priority given to afforestation in the minds of local people have all hindered tree-planting efforts (World Resources, 1986). In the face of these obstacles, suggests FAO, "small, discrete efforts building either upon skills and resources women already possess or on carefully staged, well defined inputs" prove to be most successful (FAO, 1983).

Integrated projects involving several objectives are especially valuable. An ILO/UNDP reforestation project in Cape Verde, in which up to 90 percent of women participated, included literacy classes and other training and increased participants' incomes, in addition to accomplishing its primary goal of rejuvenating the forest (Williams, 1985).

Agroforestry

Agroforestry, "the practice of combining woody perennials (trees, shrubs, palms, bamboo) with agricultural crops and sometimes animals within a unified production system" (French, 1986), can offer women many benefits. A recent ILO study notes:

> Agroforestry is a promising approach for poor households, and especially women, as it replicates the multiple products of the natural forest and can be practised intensively around homesteads and gardens, or more extensively on marginal communal lands. Agroforestry is especially relevant to women's primary concern with food cultivation, since production of wood for fuel can be usefully integrated with agricultural systems (Cecelski, 1985).

For many rural women, agroforestry concepts are traditional (Fortmann and Rocheleau, 1985). Agroforestry was, for example, long practised by Sahelian residents, and has only recently been "rediscovered" by forest development experts. The typical landscape of Sahelian countries is "farmed parkland": individually managed fields, pastures and fallows, dotted with trees. The mixture of crops and trees, such as shea-nut, nere and baobab, in many Sahelian villages, shows that certain agroforestry techniques are already being used by villagers themselves, without much outside assistance. Unfortunately, many foresters have long claimed this approach as their professional turf, while agriculturists seem to lack enthusiasm for this type of land use. Only a really integrated effort which truly aids rural farmers, particularly women, in their patterns of resource management will succeed (Williams, 1984).

To benefit from forest projects and forest management, women must not only participate, but be assured of tenure and adequate land for food production. Simple measures improve their ability to produce fuelwood from existing forest resources – providing cutting tools and transport, for example – but care must be taken to ensure these remain under women's control and are environmentally sound.

The need for an appreciation of women's role in forestry is more urgent than ever. International plans and programmes to safeguard the world's forests are being launched, the most salient of these being the FAO/UNDP's Tropical Forest Action Plan, which calls for a doubling of the annual forestry allocation to $8,000 million, and the "Silva" plea for trees to be planted on Sahelian and Sudanese drylands (Paris, 1986). The impact of these programmes on poor women and their environment must be considered with great care, and women's input must be sought. As the second part of this chapter has shown, success comes when new styles of forest management take account of women's needs and ensure their involvement in protection and planting schemes, commercial forest management, and improved social forestry and agroforestry practices. The case studies give further evidence of the good work in progress. These efforts, however, are still few and small in scale compared with the appalling and accelerating rate of forest devastation.

CASE STUDIES

THE STORY OF GADKHARKH VILLAGE, INDIA

Tucked away in one of the remoter corners of Garwal Himalaya at 1,266 metres is the sleepy village of Gadkharkh. At first glance life appears to be as harsh as in many other Himalayan villages. A steep footpath connects the village to the nearest bus station, nearly five kilometres away. The terraced village fields yield hardly enough food to provide an average family's requirement for six months, so men migrate to the cities in search of jobs. The women have to wait for money from them to buy essentials. Among the 200 members of the 25 Gadkharkh households, there are only 20 men to share all the "male" work such as ploughing the land, carrying the dead to funeral pyres, the masonry and carpentry. Women must bear the multiple burden of tending the fields and domestic animals and carrying out their normal domestic chores.

The drudgery of collecting fuel, fodder and fibre takes its toll on Gadkharkh women: their average life expectancy is 45 years compared to 52 years for men. They labour sometimes ten hours daily to collect the resources they need – primarily from a designated forest under the control of the Forestry Department. As deforestation proceeds, the struggle to find headloads grows bitter and more arduous. Blanket commercial felling of broad-leaved species such as oak, deodar and bhimal directly deprives the women of feed for their draught animals, fuel for their ovens and water in the springs.

Women take action
What is unusual about Gadkharkh is that the women have called a halt to
deforestation. They sing:

> These oak trees
> save and worship them,
> because
> their roots store water,
> their leaves have milk and fodder,
> the breeze blows cool
> over the beautiful rhododendron flower.

The Chipko movement began here in the late 1970s and succeeded in securing
a ban on cutting green trees over an altitude of 1,524 metres for a limited period
in the hill districts of Uttar Pradesh. But two problems remain. First, the
Forestry Department persisted with plantations of commercial species such as
pine, eucalyptus and poplar to replace deforested lots where oak, deodar and
burans had stood earlier. While commercial species on public land earn
revenue for the government and provide raw material for forest-based
industries in urban areas, women's requirements are still not met. The second
problem is that the Forestry Department is hopelessly ill-equipped to afforest
large tracts of deforested land and does not cover village lots which are for local
use.

Communal village pasture lands, forest lots, springs and festival grounds
have long been neglected. Their control has remained with the male Panchayat
(village council), although the rights of use and care of the village forest have
been in the hands of the women. But with the declining power of the
Panchayats, the forest fell to private exploitation and destruction by villagers.
The alienation between villagers and the Forestry Department that flowed
from their divergent interests made the women indifferent to conservation.

The Gadkharkh women's effort to afforest is important on two counts: first,
it reasserts women's rights to village forests and, secondly, it shows their deep-
rooted collective spirit, oriented to meeting people's needs.

But not only women were involved in Gadkharkh: a young man, Sachidanand Bharati, joined the Chipko movement in the late 1970s. Now a lecturer at
a local Intermediate College, he had the necessary time and energy to guide his
own people.

How it happened
A series of village meetings that started in 1982 gave birth to a women's forum
called Mahila Mangal Dal (MMD) of which all the Gadkharkh households are
members. A savings fund for women, begun with a monthly subscription of one
rupee, provides interest-free loans to its members to buy goods. They and the
fund's managers (who are all women) meet twice a month to review policy and

process loan requests. The treasurer, having been to primary school, is the only "educated" member, and the fund relies on mutual trust, for there are no written records.

By 1983 group unity had been established and the women felt bold enough to take up the issue of forest rights. There is always a tug of war between Forestry Department officials and the village women; each holds the other responsible for destroying the forests. This often results in confrontation with Forestry Department guards who are also looking for opportunities to exert their power and extract a few pennies for their own pockets. The women realized the limited value of confrontation with such a powerful bureaucracy so they turned their attention to revitalizing the Panchayat forest, which is under the village's domain.

One evening, a collective decision was taken and early the following morning eighteen women marched to the village forest with sickles in their hands and surrounded it. A man from the village was caught with a load of freshly cut wood. In deciding to punish the man, the women acted against all Indian patriarchal norms. Before the village gathering in the evening the culprit paid a fine of Rs 25. Since that incident, rotating teams of two women, sickles in hand, guard the forest every day. The women now have tremendous confidence. Gurli Devi, president of MMD, says: "When men governed the forest, it was destroyed: therefore, we have taken the responsibility for protecting the jungle into our own hands."

Village self-help
A small nursery for young saplings has been set up on land donated by villagers and broad-leaved species are being planted, including walnut, bhimal and pangar. Twice a year, hundreds of young students, villagers and social workers get together at the plantation camps, not only to discuss environmental issues, but also to dig holes in rocky land for planting young saplings. The participants bring most of their own food and the villagers provide accommodation, so the cost of these plantation camps is low. The mobilization of voluntary labour has also made it possible to build stone walls around the village forest and the new plantations.

Initially, the MMD strategy was to protect the remaining forest – there were no resources to carry out large-scale afforestation on the 70 hectares of village land. The plantation camps have, however, enabled them to replant as well.

After four years, the village lot is now open for the removal of fodder grass and matured trees to meet the domestic needs of those families recommended by the MMD. Its members use energy-efficient, smokeless ovens and pressure cookers, although there is not yet a subsidy for these. Gadkharkh is helping 25 other villages in the area to emulate its example and revitalize their own forests.

The lessons from Gadkharkh

The Gadkharkh women's spontaneous initiative to guard, protect and regenerate the forest came from their experience that conservation guarantees survival. They argued that a forest's success is not only measured in terms of the profitability of its trees, but in its capacity to fulfil women's needs. All that was needed was a catalyst, in this case Bharati, to spark their spontaneity. Illiterate and overworked these women may be, but they have systematically organized the management of the forest.

The example of Gadkharkh shows clearly that tree planting requires shared responsibility for the land and the forest by the people of the village. Each sapling in the Gadkharkh plantations is nurtured with great care and tenderness by people who have control over the product. The homogeneity of the village, which has no caste or class barriers, also played an important role in making this programme successful.

Now, the need of the hour is that appropriate hill and household technology be introduced to lessen the women's daily drudgery. Women themselves must be involved in such innovations because they have valuable suggestions and solutions to offer. "Women have more discipline in carrying out any movement," says Gurli Devi. "Women can take it till the end, men change their minds."

Source: Renu Wadehra, India.

THE GHOREPANI PROJECT, NEPAL

Nepal's rural people, who make up over three-quarters of the population, have a crisis on their hands. In their efforts to satisfy the increasing need for food and fuel, they are stripping steep and unstable slopes for firewood and fodder, and clearing and overgrazing the pastures. Much fertile soil is being lost through erosion and landslides. If present rates continue, Nepal's forests will disappear within the next fifteen years. Women bear most of the burden of this environmental degradation.

Since Nepal opened its borders to foreign visitors in 1950, the dramatic increase in trekking has also damaged the natural and cultural environment. Ghorepani village, in the hills of Nepal, is a prime example. Tourists' demands for food and services have aggravated deforestation and had a major impact on the social and economic life of women. In addition to their traditional duties of fuel, fodder and water collection and agricultural labour, women and young girls perform many of the tasks involved in servicing the tourist industry, such as cooking and cleaning for trekkers. It is tourism (among other factors) that has discouraged girls from continuing education, as their labour is perceived to be of more value than that of boys. And some women have forgone marriage, since they feel a husband and children would only add to their existing work burden. Tourism has created some employment in tourist lodges, but the girls are rarely paid and earn only their keep.

The Passive Environmental Development Centre
With financial support from the Australian Development Bureau, the Australian Association for Research and Environmental Aid (AREA) started a project in the village of Ghorepani. The aim of the Passive Environmental Development Centre is to relieve the immediate human pressures on the environment to reach a state capable of sustainable development. The project has many elements, including:

- the development of fuel-saving energy systems, such as improved heating and cooking stoves, solar cooking and heating systems and the improved insulation of buildings;
- upgrading of domestic water supply and the use of natural methods of water purification;
- the introduction of appropriate sanitation systems;
- a community forestry programme which encourages sustainable-yield forest harvesting and the rehabilitation of degraded land;
- the use of innovative methods to encourage the sustainable production of local materials;
- agricultural assistance to improve efficiency and sustainable methods of cultivation;
- the improvement of livestock management;
- a programme of project participation for women; and
- a continuous programme of environmental monitoring.

The project is small but flexible with maximum community participation. It is based on developing skills among the village tradesmen and the local community while encouraging greater environmental awareness.

Women's participation programme
The appointment of a Women's Liaison Officer to the project was considered essential if the programme was to gain maximum acceptance and participation and be of benefit to the whole community. Her role has been to encourage village women to join in the decision-making and assess the suitability of new technologies.

Together, they have helped to identify the lodges and households wishing to have passive energy, sanitation and water supply systems installed and have worked with women to ensure that new systems were understood and culturally acceptable. They have tried to ensure that any changes introduced do not disadvantage women or increase their workload.

Getting women to articulate their needs was an important stage in the planning process. It was found that they were more concerned about a constant flow of water than with its contamination problems. They wanted better sanitation. Many were concerned about the dwindling supplies of firewood and fodder, and they knew that loss of forest cover in the Ghorepani area could lead to severe erosion.

Women not only discussed these problems, they worked to solve them. They did much of the excavation for the new water supply system, while both men and women participated in tree planting.

Importance of discussion

Before introducing appropriate sanitation systems, the Women's Liaison Officer discussed both the advantages of improved sanitation, particularly for children's health, and the connection between the control of flies and the spread of disease. This was done with the aid of stylized drawings, a method also used when discussing the benefits of other appropriate technologies.

Detailed consultations were necessary before many changes could be introduced. New hot-water heaters, for example, required discussion to determine whether such a system would be culturally acceptable and of benefit to the women. Women soon realized that the heaters would use less fuel wood, give them more effective control over the quantity of hot water used, and reduce their workload.

With the introduction of solar technology a different approach was taken. Demonstration units for hot-water heating and solar cooking were set up and used by the project team. The demonstrations generated considerable interest, particularly in the solar oven. Reaction ranged from initial disbelief (that food could be cooked this way) to a situation where villagers collected the silver wrappings from cigarette packets and chocolate bars to build their own solar ovens.

The original project proposed the appointment of a local woman as a counterpart to the Women's Liaison Officer. But for a number of reasons this did not happen. Women in Ghorepani have little, if any, free time, and to single out one woman would have increased her workload since she would still have had to fulfil her other domestic responsibilities. And among Magar Pun women (the predominant caste) there are strict kinship and neighbourhood ties which follow a strong hierarchical order but which also form the basis for working groups. So, in practice, it was found that working with existing "women's groups" was a more effective way of spreading information quickly, and one which did not threaten the existing social structure.

The establishment of a rapport between the Women's Liaison Officer and the community, particularly the women, was achieved by her willingness to learn from the women and to perform the same domestic duties as they did. Men also accepted her because she followed a pattern of expected female behaviour which did not confuse or threaten them.

Because the project is small and targeted at a single community, its interrelated objectives – to gain maximum environmental protection while allowing sustainable development – are being achieved. Flexibility is the keynote of the project. As no two environmental or cultural situations are the same, technologies must be adapted to suit individual situations and needs.

Source: Josette Wunder, Australian Association for Research and Environmental Aid.

WOMEN IN FORESTRY: THE KENYAN CASE

Kenyan women have been in the forefront of forest management for many years. As food and fuelwood gatherers, they have developed close contacts with nature and a dependence on its abundance. Many women (and men) collect medicinal plants, and the raw materials for goods such as baskets and mats are also often supplied from forests.

The women of Kenya understand their vulnerability when the environment is degraded: a large number of women's groups are involved in tree planting and soil conservation.

Historically, Kenyan women have played a part in establishing the exotic wood plantations which form the basis of the country's wood industry today. Figure 4 shows how they were employed in 1946. The high numbers of women and children are explained partly by the exploitative policies of that time and partly by the Shamba system.

Figure 4: *Forest Labour Census of 1946*

	Resident Labour Employed by the Forest Department	Labour Employed by Forest Industries	Total
Men	4,091	3,738	7,829
Children	11,537	7,779	19,316
Women	5,829	4,530	10,359
Total	21,457	16,047	37,504

The Shamba system

This form of shifting cultivation supplied the Forest Department with cheap labour for tree planting while providing the labourer with a plot of land on which to grow food. Families living in the forest were allocated a piece of land which they cleared and then planted with annual crops such as maize, beans and potatoes for one or two years. The Forest Department then moved in and planted trees. These were cared for by the labourers, who continued to plant food crops between the rows until the tree canopy closed. The labourer families were then allocated new plots. Women and children did most of the work, for men were often engaged elsewhere.

The Shamba system was abolished in the 1970s and replaced by a modified scheme in which land plots are hired out to any willing cultivator. Even today, however, most of the cultivation is done by women who are left behind in the rural areas while men flock to the urban centres in search of paid employment. The Shamba system can be seen as a forerunner of modern agroforestry in which agricultural crops are deliberately grown on the same piece of land as trees.

Women in forestry
Although women have played a major role in tree planting, when it comes to paid employment they are left mainly with the menial and less well paid jobs. Women are to be found making secondary forest products such as strawberry punnets, cheese boxes and safety matches. Their dexterity assures them a job in the packaging sections of most industries but rarely are they involved in management, ownership or professional positions.

In forestry, operations such as picking out tree seedlings from germination beds is a job thought to be well suited to women. It is said that their patience and nursing instincts ensure that the delicate seedling, usually only two to three weeks old and bearing the first two leaves, is carefully uprooted and moved to another bed, or to polythene tubes in which it grows to the required size before it is transplanted to the field. Women also do other nursery work, including filling polythene tubes with soil. But, again, few women will be found at the policy- and decision-making levels in government or the private sector.

Figure 5: *Positions in Kenya Forest Department, 10 July 1986*

Job Title	Male	Female
Chief Conservator	1	–
Deputy Chief Conservator of Forests	1	–
Assistant Chief Conservator of Forests	1	–
Conservator of Forests	18	–
Assistant Conservator of Forests	79	4
Foresters	307	5
Forest Assistants	154	–
Total	561	9

The first Kenyan woman forest officer was appointed in 1971, 69 years after the Forest Department began in 1902. But that one advance did not open the doors for other women, as figures from the Kenya Forest Department show.

Conservators of forests and the grades above are the main decision and policy-makers. All of these are men, so it is not surprising that matters of importance to women are given a low priority. Forest planners have ensured that enough timber is supplied for building and industry, but they have, until recently, failed to plan for firewood needs. Nor have minor forest products such as mushrooms, fibres for cottage industries, medicinal plants, root and leaf extracts featured anywhere in forest planning, although they are important for rural women. Moreover, the lack of interest in what Kenyan forests can provide resulted in policies that accord low priority to indigenous species in favour of "fast-growing" exotics – even though it is known that exotics suffer from disease and pest attacks to which indigenous trees are resistant.

The absence of women at policy-making levels is not merely because of their exclusion by planners. It is partly a result of conditioning, and the fact that boys have more educational opportunities than girls. Although theoretically there is no discrimination in job opportunities, certain jobs are universally regarded as a male preserve, and forestry is one. The type of forestry traditionally practised in Kenya (and indeed in many other countries) has made it difficult for women to join the profession, for they must work in remote areas, often policing huge, government-owned forests. Job opportunities were thus limited to research services which, until recently, were not well developed.

New opportunities
In 1971, the Kenya government established a Rural Afforestation Extension Programme with the aim of taking forestry to the people. It is generally recognized that women make good extension workers. Now they can serve in this capacity and work with the numerous international organizations and NGOs involved in tree planting. Research services have also been reorganized and offer better opportunities. And women who want to influence policy- and decision-making at the national level can find jobs in education. However, all these areas require highly specialized training and dedication: women have yet to take the challenge and make the crucial impact they can in forest management.

Source: Theresa C. Aloo, Forest Department, Egerton College, Kenya.

CHAPTER FIVE

Women's Energy Crisis

Energy, although still not officially considered as a basic human need, is essential for human well-being and forms the driving power of economic development. Within households, energy is needed to cook food, to boil water and to heat and light the home. These are women's tasks, and this chapter shows how the energy sector links environmental conditions directly with the situation of women.

ENERGY: A BIOMASS CRISIS

Chipko marchers went through the Himalayas in 1982. They met an ex-soldier in Bhatadi near Mahendranagar in Nepal:

> They asked him: How do you meet your fuel requirements?
> The answer was: There was no problem in the beginning. We had enough twigs and branches. Then we shifted to the dust and refuse of saw mills. Then we used stumps. We are now going to dig the roots.
> Asking again: And then?
> No reply.
> The reply was available to the marchers, the writings on the wall were the cow dung cakes (Ummaya *et al.*, 1983:82).

Most domestic energy in the Third World comes directly from biomass sources. Woodfuels (both firewood and charcoal) and other biofuels, such as animal and crop residues, are the main sources of energy for some 2,500 million people, or roughly half the world's population. It is true that, in spite of price increases, oil consumption has risen and will continue to grow in developing countries. And electricity production, especially from hydro sources, is expected to increase (World Resources, 1986). But these commercial energy sources do not often reach the poor – even in the city. Electricity may be available, but not used: an ILO study on rural Indonesia showed that the rural poor could not afford to buy electricity (Puerbo, 1985).

In some countries, charcoal and wood remain the predominant

cooking fuels (even for the middle classes) and this pattern is repeated throughout the Third World. Overall, more than half the total energy consumption of poorer countries is believed to consist of traditional fuels, especially wood (Cecelski, 1986). Nepalese families depend on wood for 97 per cent of their household energy; in some African countries – Burkina Faso, Chad, Ethiopia, Malawi, Mali, Somalia and Tanzania, for example – wood accounts for 90 per cent of national energy consumption. Even in oil-rich Nigeria it accounts for over 80 per cent of energy consumption (Eckholm *et al.*, 1984; *Courier*, 1986).

Causes of the crisis

Firewood collection for household use is often blamed for deforestation; but normally, rural people use only dead wood, for it is easier to cut, lighter to carry and burns better than green material. Branches may be cut, but whole trees are rarely felled merely to provide fuel (World Resources, 1986). On a global scale it is urban, rather than rural, consumption of wood that may be more responsible for depleting rural wood supplies (Manibog, 1984).

Fuelwood is carried or trucked from rural areas to many towns and cities. Studies have shown that around Ouagadougou (Burkina Faso), Dakar (Senegal) and Niamey (Niger), a deforested "ring of desolation" is the result of this traffic. Wood supplies come from more than 100 kilometres away, with charcoal brought in from even greater distances. In refugee camps, such as those of Pakistan, Somalia and Ethiopia, and collective villages in, for example, Mozambique and Tanzania, fuelwood is a major energy source.

Locally, rural industries deplete the growing stock. Rural industries and agricultural processing outside the household use large quantities of wood for energy: 90 per cent of Tanzania's industrial fuel comes from wood. Some processes consume large quantities: in Malawi, 3 to 12 kilograms of tobacco require 100 kilograms of firewood; the same amount is needed to bake 50 kilograms of bread in Guatemala or to smoke 66 kilograms of fresh fish in the Ivory Coast (*Courier*, 1986). And although charcoal production requires considerable forest resources it is often a last resort for poor families who need food (Cecelski, 1985-1). Around the Hargeisa camps in Somalia, where there are few other opportunities for income-earning, some 2,000 refugees have taken up charcoal production for sale (Hayes, 1982).

With the depletion of fuelwood supplies, poor people are forced to switch to other cheap biomass substitutes, such as straw, crop residues and animal dung. Worldwide about 800 million people now rely upon residues for their energy needs (Barnard *et al.*, 1985). The tragedy of this dependency on wood is that it is being depleted more rapidly than any fossil fuel, and that its consumers have little political power. A 1983 FAO study showed that more than 100 million people suffer an acute fuelwood scarcity and are consuming amounts below the minimum required for cooking and heating. Shortages are most acute in the arid regions of Africa, the mountainous areas of Asia, and the Andean plateau of Latin America (World Resources, 1986). Projections for the year 2000 show that without immediate action to improve the situation, 3,000 million people will either be unable to obtain minimum energy needs or be forced to consume wood faster than it is being grown (Barnard *et al.*, 1985). Although the data for these studies are often poor, the trends are clear. The energy crisis of the poor is serious and worsening every year.

Consequences for women

> The boys have nothing to do. For girls it's different. We have to look after the family collecting firewood as before, but we have to walk further each day to find any (Sara, an Ethiopian refugee of 14 living in a Somalian camp).

The burden of the fuelwood crisis is borne by women because they have the responsibility for meeting household energy needs through fuel collection, preparation and use. Children in almost all developing societies must help their mothers with these tasks. Girls especially take part in fuel preparation, cooking and tending the fire. In Tanzania, young girls begin to help their mothers as soon as they can walk (Cecelski, 1985-1); in parts of Africa, mothers take their daughters from school to help them gather fuel (Hoskins, 1979).

Men may share the task, but in most countries, especially in Africa, women are responsible for fuel collection. Women in Ghana traditionally collect all the household fuel, while men work in charcoal production (Ardayfio, 1986). Men in Rwanda occasionally cut down trees from their wooded land and bring large logs to the household, but the daily work of gathering dead wood, bushes and banana leaves, and cutting and transporting eucalyptus offcuts is done by women and children (Bart, 1980). When it is a distant forest in Nepal, with heavy timber that requires chopping with axes, men usually do this, but women go out daily to collect brushwood and tumblewood from the hillsides (Schuler, 1981). In India, men in the Uttar Pradesh hills are found to break the traditional division of labour only by fetching fuel and fodder when the productivity of women's labour is high, for example on irrigated land (Srinavasan, 1984). In Peru, men and women share more equally in fuel collection (Alcantara, 1985).

As domestic fuel becomes more commercialized and collection is oriented towards large-scale organized sale and charcoal making, men's participation increases. But so long as technology and marketing are absent, the task of fuel gathering is relegated to women. Women lack material and social resources to compete with men, for they seldom have access to donkeys, carts or trucks for transport, or the tools and technology for felling and charcoal-making. In Kwenzitu, Tanzania, every aspect of fire tending and fuel gathering is the work of women; no task is considered to be as tiring or demanding, or to have so little to show for itself (Fleuret, 1978).

Throughout the world, women are carrying loads up to 35 kilograms over distances as much as ten kilometres from home. The weight exceeds the maximum weights permissible by law in many countries, which typically prohibit manual carrying by women of loads greater than 20 kilograms (ILO, 1966). These heavy burdens damage the spine and cause problems with childbearing. The back-breaking work of collecting, cutting and transporting wood, exacerbated by poor nutrition, undermines women's health still further. And the longer they have to walk, the more they are affected.

As wood supplies become scarce, women must walk substantially longer distances and spend more time and energy in fetching firewood. The real energy crisis is a lack of women's time, concludes Irene Tinker in a study on rural energy and the position of women (Tinker, 1984). In the foothills of the Himalayas, just a generation ago the gathering of firewood and fodder took no more than two hours; now it takes a whole day and involves walking through difficult mountainous terrain. Over the last decade, the time taken to collect firewood in parts of the Sudan has increased more than fourfold (Agarwal, 1986). And on the Gujarat plains of India, where fuelwood was traditionally collected once every four days, its depletion means that four to five hours *per day* are now spent collecting in some areas (Nagbrahman *et al.*, 1983). Up to ten per cent of Peruvian women's time is spent in fuelwood collection (Foster, 1986), while in Gambia, women spend from midday to nightfall every day gathering the evening's supply.

Fuel preparation (the chopping and drying), cooking and care of the fire are almost exclusively tasks for women and young girls, and this also takes many hours each day. In the Peruvian highlands, for example, where women spend about one hour a day gathering fuel, it takes four times as long to cook the meals (Skar, 1982).

As women's working days are already very full, increasing the time spent on firewood collection and cooking becomes an overwhelming burden. Less time is left for more productive, income-generating activities (Cecelski, 1985-2).

Not only must women face the depletion of wood from within areas where they traditionally collect, their access to those areas is declining as they lose control over land and its resources. Rural landholders can gather firewood and crop residues from their own property. The landless must depend on wood from common lands or may be allowed to gather it from other people's land in exchange for their labour (Agarwal, 1986). The growing inequity in land-ownership patterns results in high inequalities of fuel availability.

Women in particular have poor access to land of good quality and have no guaranteed long-term rights. Trees on household lands are often considered men's property (Cecelski, 1985-2), which they sell as wood or charcoal as firewood becomes scarce and prices rise. Women, denied an income, cannot afford to buy fuel.

Less fuel: more problems
When not enough fuelwood is available, women shift to alternative fuels, such as cattle dung and crop residues, coconut husks, rice hulls, millet stalks, or herbs. In some countries, such as Bangladesh, where all the fuelwood has gone, these are the only "free" fuels left (Barnard, 1985). Gujarat women depend increasingly on tree roots, weeds and shrubs and, at times, even on straw. But these fuels, although more abundantly available, are less convenient for cooking. They may take less time to collect, but they make more work later for the woman, who must continually feed and tend the fire. Cooking requires more of these inferior fuels and takes longer. And the smoke coming from weed-like bushes, such as retama in Peru and basoothe in India, is often even more poisonous than that of fuelwood (Cecelski, 1985-1).

The use of dung and crop residues as fuel seriously affects land fertility, reducing soil and mineral retention. This causes a loss of agricultural output. If the dung had been used as manure, agricultural yields would have been higher. Every ton of cattle dung burned means a loss of approximately 50 kilograms of food grains. In the Third World about 400 million tonnes of cattle dung are burned annually. That means a loss of 20 million tonnes of grain each year (Spears, 1978). Dung as well as crop residues are linchpins of the agricultural system, and their increased use for fuel could have devastating effects (Eckholm *et al.*, 1984). The loss of agricultural productivity again places a heavier burden upon women.

Fewer cooked meals
Fuel and food are complementary needs. Most foods, particularly the wholegrain and legume staples on which most rural diets are based, are inedible without some cooking. A close statistical association exists between per capita consumption of food and that of fuel (Cecelski, 1985-1). In Sri Lanka lower-income groups eat foods which require little cooking, whereas wealthier families consume higher-priced and more fuel-intensive rice and hot curries.

If less fuel is available, often a change in cooking habits takes place. In many Third World countries two cooked meals a day are usual. But in several regions a reduction in the number of meals cooked is now taking place because of fuel shortages. In Bangladesh and the Sahel, a shift from two meals to one a day has been observed (Agarwal, 1986). And in Korea, the shortage of ricestraw from crop changes has reduced cooking to one meal a day. In densely-populated Rwanda, 62 per cent of families cook only once a day and 33 per cent cook even less frequently. The number of cooked meals is relevant for the health of the family and particularly critical for young children, who need at least three small cooked meals a day.

Fuel shortages are also reflected in the shift to less nutritious foods (which need less energy to cook) and in a reduction of traditional foods. In the highlands of Mexico, beans have been the principal source of protein for the poor, but they require an enormous quantity of wood (15 kilograms of wood per kilogram of beans). This means that the poorer the family, the fewer beans are eaten (De Cuanalo, 1983). Similarly, in Guatemala a shift from the staple diet of beans to other less fuel-consuming foods is taking place. And in the Sahel, millet has been replaced by rice. Attempts by authorities to introduce soybeans in Burkina Faso have failed because they require a longer cooking time and a greater quantity of fuel than the traditional cowpeas (Hoskins, 1979). In many places, a shift to more raw food or partially cooked or cold leftovers is becoming commonplace.

It is difficult to distinguish the effects of fuel shortages from the effects of lack of time and of food itself on cooking patterns, especially in African countries with severe food shortages. But the lack of appropriate fuels clearly contributes to inadequate nutrition. Partial cooking also causes serious health problems from infection. Women in many communities are the last to eat, often taking only the leftovers from the family meal. Not only is their nutritious intake limited, but they also risk infections from rotting food. Water purification by boiling, while not widely practised, also requires fuel. Hot water is needed for washing cloth, cooking utensils and people. These are essential requirements for maintaining a germ-free environment and preventing the spread of disease. As fuel supply diminishes, all these basic health practices are curtailed (Cecelski, 1985-1).

Space heating is often necessary not only for health but for survival, particularly for families living in mountainous areas and on high plateaux. These high-altitude regions are not uncommon in developing countries; they are often – as in Ethiopia, the Himalayas and the Andes – environmentally-stressed and fuel-short areas as well.

Indoor pollution

Emissions from biomass fuels are dangerous sources of air pollution in the home, where women cook during all or part of the year (WHO, 1984). Wood fuels are capable of producing pollution concentrations higher than fossil fuels under slow-burning conditions and some studies have shown that cooks suffer from more smoke and pollutants than residents of the dirtiest urban environments. They are affected by a higher dose than is acceptable under the World Health Organization's recommended level or any national public standard (Smith *et al.*, 1983). In one study quoted by the World Health Organization, a female cook can inhale an amount of benzopyrene (a poisonous gas from burning fuel) equivalent to 20 packs of cigarettes a day. In a few places chronic carbon monoxide poisoning is also evident (WHO, 1984).

It seems likely that respiratory and eye diseases, which are so abundant among Third World women and children, are caused by wood and other biomass burning. Exposure can bring acute bronchitis, pneumonia and death where respiratory defences are impaired. Studies also show that where emissions contain high concentrations of carcinogens, nasopharyngeal cancer is common among young people who have been exposed since infancy (WHO, 1984).

Fuelwood and income

In India, the Renchi firewood sellers, mainly tribal women, usually begin their day at 2.00 a.m. in order to complete household chores before setting out on the 8 to 10 kilometre walk to surrounding forests where they collect wood. (Seven or eight years ago, the forest was just a kilometre or two away.) The wood is then sold in the nearest town, Renchi. To reach the market early, the women must leave their villages the previous evening by train or bus and spend the night at the railway station or some other public place. The headload they carry sells for Rs5.50 to Rs6.50. About a third of this money is lost in bribes to forest guards, bus and train conductors. ... Work begins again as soon as they return home, in cooking the evening meal. The next morning they again set off early to collect wood (State of India's Environment Report, 1984-85).

Headloading – the collection of firewood, carrying it and selling it elsewhere – has become an important activity for women in several countries, including India. It began in Bihar after a drought left households destitute many years ago. And the practice is spreading in other regions of India, the Sudan and elsewhere. With the growing scarcity of fuelwood, its commercialization is increasing, mainly in urban areas where prices have more than doubled in the past decade. In cities such as Ouagadougou, Burkina Faso, and Port-au-Prince, Haiti, urban families must pay up to 30 per cent of their meagre incomes for cooking fuel (Foster, 1986). It seems that the impacts of the commercial fuel trade are spreading outward from the cities. In Malawi only 7 per cent of rural families buy any fuelwood, but 65 per cent of households in a village in South Yemen get at least a quarter of their fuelwood from the market (Eckholm *et al.*, 1984).

Although commercialization means an increased income for some, it also places an extra demand on family budgets, posing special dangers for the poor. In parts of India, landless families who once had the right to collect wood freely on private land have lost this privilege. Landowners now see a chance to sell their excess wood for cash. Women, again, are the first to suffer.

But women, too, often rely on wood products for an income. In countries which formed part of the ILO study on energy and rural women's work, women's contributions to household cash income varied from one-third in some Indian villages to nearly 80 per cent (during the dry season) in Ghana (Fleuret, 1978).

Fuel is a key raw material for most of women's typical income-generating activities, such as food processing, beer brewing, and pottery. All the major crafts in Mali, except mat-making, are fuel-intensive (Koenig, 1984). In many West African countries, women earn significant incomes by producing dolo beer. In Botianor, a coastal village of Ghana, one-third of women's income comes from the sale of smoked fish (Ardayfio, 1984). But as fuel availability decreases and the cost rises, these activities become less viable and sometimes they are abandoned. Between 1977 and 1981 wood prices in Ouagadougou rose nearly 28 per cent annually, while the selling price of dolo rose more slowly, decreasing incomes (Gattegno *et al.*, 1983). In Ashele Botwe, Ghana, nearly every family (and particularly women) made charcoal (see Afi's story, page 80). But they quickly exhausted the wood supply on their own fallow lands, and access to trees on other private lands became impossibly expensive (Ardayfio, 1984).

COOKING AND THE FUELWOOD CRISIS

There are two basic solutions to the rural energy crisis: increasing wood supplies, and improving cooking efficiency. Chapter 4 discussed some of the ways in which wood supplies could be increased through better management, reforestation, and social and agroforestry. The rest of this chapter looks at various programmes concerned with improving the efficiency of cooking.

In most rural areas in developing countries, particularly those in Africa, women cook on three stones or on primitive stoves. The efficiency (that is, the caloric value which is transformed into useful cooking energy) of primitive stoves is usually less than 10 per cent. The "total efficiency", including fuel and non-fuel factors, could be much higher (Cecelski, 1985/1986); simple improvements, for example a mud enclosure around an open fire to protect it from draughts, can improve efficiency considerably (Eckholm *et al.*, 1984).

In India, efforts to improve the traditional method of cooking began more than 25 years ago. The earliest positive effort is found in the stove designed during the time of Mahatma Gandhi, called the Magan Chulha ("chulha" means stove) (Gupta, 1985). Since then, a whole range of improved stove programmes has been developed, yet many have failed. Evidence suggests that improvements in other factors, such as management of the cooking fire, utensils and the way food is prepared and processed, can save as much fuel as improved stoves.

Some improved stoves require only a marginal change in people's cooking habits; others necessitate a complete reorientation. Low-cost traditional wood-burning stoves can often be improved through minor engineering changes. These stoves have many different shapes and sizes, are frequently made from locally available materials such as mud, and are often portable, though proper maintenance of dampers and chimneys made of sheet-metal is critical to their performance (Cecelski, 1985-1). Adoption of traditional mud and ceramic stoves is already widespread in Asia (Eckholm *et al.*, 1984).

Some experiments designed to improve cooking efficiency have overlooked the importance of traditional cooking habits. Most charcoal- and biomass-burning stoves are portable and made of sheet-metal or ceramic materials. Biomass availability, often seasonally based, is a determining factor in their use. Charcoal, on the other hand, usually has to be purchased, which makes the use of these stoves out of reach of the rural poor. Hay boxes or heat cookers are simply well-isolated boxes with a lid. They use less fuel, and are cheap and safe. But they require a major change in cooking methods. Although solar ovens can in theory be of great use, in practice they have not yet been adopted on a large scale. Most are of relatively high cost, difficult to operate and require that cooking take place outdoors and during the daytime.

Biogas plants use animal and sometimes human and crop wastes, which are anaerobically fermented. They produce high-quality cooking gas and slurry, a very good fertilizer. In India and China biogas stoves have been promoted widely for household cooking, but their costs are high, and access to sufficient animal dung and other residues is necessary. For poor families – who normally do not own more than one or two animals – these are insurmountable obstacles. Community biogas plants have had even less success because of considerable organizational problems in waste collection, plant management and the distribution of benefits.

Women's role in effective stove programmes
Several factors have influenced the spread of more efficient stoves: their design and development, their cost, the need for infrastructural change (including extension services and credit facilities), and the attitudes and status of women (Agarwal, 1986).

It is clear that improved stoves have not been widely accepted. An evaluation of programmes in Gujarat, India, demonstrated that sponsoring agencies assess their success primarily in terms of the number of stoves which have been built. In no case had the sponsoring agency tested whether the stoves built were less smoky or more efficient in practice than traditional fireplaces (Skutch, 1985). Little is known about the sustained performance of new stoves, for most surveys are carried out soon after they have been installed. Many stoves use, under field conditions, almost as much fuel as an open fire; laboratory testing alone cannot replicate the field situation. There is often almost no flexibility possible in cooking methods: only pots of a certain shape can be used. Factors such as time saving, convenience, smokelessness, house heating and food testing are at least as important for the users as is fuel efficiency. Improved stove designs often fail to offer these other qualities.

Perhaps the most important reason why many programmes have failed is that women are reluctant to adopt stoves or participate in schemes where they have made little input to the design. The typical procedure is that male technicians design laboratory-efficient stoves. Then female extension workers demonstrate them and persuade the potential female users. This means that local women's experience, knowledge and needs are not taken into account. In an Afghan refugee camp in Pakistan, for example, male experts designed improved mud stoves without noticing that women themselves were already building these enclosed mud stoves, complete with chimneys and space for water heating.

Any stove programme should include women from the start, as planners and designers, builders and promoters. Such a programme should begin with a survey of the local situation and user needs. Action research of this kind, as used in the ILO Energy and Rural Women's Work programme, can provide much necessary information for analysis (Cecelski, 1985-1). The stove model finally introduced must be one that local entrepreneurs or users themselves are willing to build without outside assistance (Eckholm et al., 1984).

Programmes and projects
Such an approach has been followed by an ILO/World Bank Stoves Feasibility Project in Ethiopia, which focused on urban users in Addis Ababa. First a survey was made of characteristics of the households using different stoves, appliances and fuel types. In collaboration with neighbourhood women's associations, prototypes were then constructed. Different methods of marketing and dissemination, such as the sale of stoves through women's groups, were also tested. On the basis of these results, stove and pot production and promotion strategies were developed. Eventually a plan for large-scale dissemination was prepared (Gebreselassie, 1985). This project showed how women, the end users, can be involved in a programme from the start.

Sixty NGOs met in Kenya on the occasion of the UN Conference on New and Renewable Sources of Energy in 1981, and in 1982 they officially established the coalition KENGO, the Kenya Energy Non-Governmental Organizations. Among the member-NGOs are women's groups, church and youth groups and international development agencies. KENGO, headed by a man, has a professional staff of six women and 14 men. They execute a number of programmes, of which the indigenous trees programme, mentioned in Chapter 4, is one.

KENGO's Wood-Energy Programme includes a number of domestic and institutional cookstove development projects, including the Kenya Ceramic Jiko and the Kuni Mbili Stove. Both are based on designs tested by women in the field. And women themselves can construct the Jiko. The institutional cookstoves project, which began in 1985, focuses on the use of firewood by schools, hospitals, bakeries and other institutions. In addition to stove development, technical assistance (particularly on agroforestry), public information and education are all important aspects of KENGO's work. REECA, the Regional Energy Environment Conservation Association, facilitates coordination on energy and environment issues in Eastern Africa, including Ethiopia, Lesotho, Madagascar, Somalia, Sudan, Tanzania and Uganda (Khamati, 1985; Musumba, 1985).

In Uganda, with rapid deforestation and population growth, firewood – once plentiful – is becoming scarce. Even fruit trees like mangoes are being cut down in some parts of the country for firewood or the preparation of charcoal. Since 1983 the National Joint Energy and Environment Project (JEEP) has been engaged in education work to promote public awareness of the energy crisis. It broadcasts a daily short message on radio and television, promotes agroforestry at the village level, and improves stoves in urban areas (Muzira, 1985).

The Sudanese Renewable Energy Project (SREP) of the National Council of Research has worked extensively on charcoal stoves, fuelwood production, solar energy and the promotion of solar cookers. These efforts have been complicated by the fact that men buy the stoves, while women operate them; to reach women more directly, SREP has tried to demonstrate the stove through women's groups, as well as through television, radio and the press (Elnur, 1985).

In Indonesia the Yayasan Dian Desa (YDD), an NGO that works on appropriate technology, has developed a stove made from fired ceramic, plastered with a clay-sand mixture as insulation. The stoves are locally produced, so that their design can easily be controlled by village users. The project began in 1983, and by mid-1985 20,000 of the stoves had been distributed in eight provinces of Indonesia. The project involves the community, particularly women, as partners in action (Wisanti, 1985).

There is evidence from India that owner- or mason-built stoves, especially those made by women, seem to be more successful than those built by outsiders (Sarin, 1983). Women are often more effective in maintenance and repair if they are able to receive training (Elmendorf, 1980). The positive results of a participatory approach are also demonstrated by the experience in Senegal, where the participation in a group activity such as stove building can have a tremendous impact on village women: they are amazed that they can do it themselves. Even if Ban-ak-Suuf stoves have something of a short life, all the preliminary work, organization, and participation that lead to the actual construction are accomplishments which contribute to community – and women's – development (Foley and Moss, 1983).

Changing food habits

Less fuel-consuming cooking methods can save as much or more energy than improved stoves. Many traditional means of food preparation, such as the soaking of beans, already do this. Improvements are possible in the management of cooking fires, utensils, and in food preparation. Access to improved cooking pots, to better tools to cut wood, and to better methods of drying and storing wood can all help. Dry wood, of course, burns much more efficiently than wet or green wood; in Peru, the Club of Madres constructed simple storage structures to keep fuel dry during the rainy season.

Food-processing activities in the household are major fuel users. In Bangladesh, for example, 20 per cent of energy is used for parboiling paddy. In Mexico, the costs of preparing tortillas at home are greater than buying them (De Cuanalo, 1983). That is why community food-processing bakeries, such as those in Pakistan and Peru, can be important fuel savers as well as important sources of income for local people.

According to Paula Williams,

> There has been a considerable amount of debate as to how significant an impact cookstove programmes have had, or potentially could have, on mitigating environmental problems. Firewood consumption for cooking is only a small part of the overall process of deforestation. Although reduction of this consumption may be relatively insignificant on a larger scale, it can have a positive impact at the individual household level.

But cookstove programmes have a wider significance than the reduction of firewood consumption. "These programmes offer women a chance to participate in development programmes, both as professionals and as users of the introduced technology" (Williams, 1984).

ALTERNATIVE ENERGY RESOURCES

Alternative energy sources such as biogas, the sun and the wind offer considerable potential for community services – water pumping and lighting, and power for small enterprises, for example. But for the poor, the capital costs of these technologies are far too high. Access to and maintenance of the systems are often complex, requiring well-functioning social structures. And women are easily excluded from these technologies. While women and girls in a number of countries are the primary collectors of dung, and the chief end-users of biogas energy, they have been only marginally involved in the planning and implementation of biogas technologies, except where women's groups are strong, as in the case of the Women's Revolutionary Socialist Movement (WRSM) in Guyana (see page 83).

Integrated energy planning
Little emphasis is given to the energy needs of the poor, in either urban or rural areas. Rural electrification has lately received great attention and costs millions, but it has failed to encourage the promised economic development (Cecelski, 1985-2). All too often, energy developments are aimed at producing power for industrial or rich urban markets alone; they benefit only large asset owners and they ignore the needs of women.

A good example of an integrated energy project is promoted by the Centro de Estudios Mesoaméricano sobre Tecnología Apropriada (CEMAT) in Guatemala. The Lorena stove, a high-mass mud stove which uses human waste, was developed there, together with a dry alkaline family fertilizer latrine, which supplies the organic material, to improve the environmental and working conditions of women exposed to smoke and to improve sanitation. These complementary technologies were combined with the promotion of mixed family gardens, in which the recycling of ash and faeces are promoted. Intercropping and the use of organic fertilizers are central. Using local materials and skills, the Lorena stove has been accepted in many parts of Central America, where it has been introduced after a proper evaluation of needs and involvement of the users (Caceras, 1986; Cecelski, 1986).

The expressed needs of existing grassroots organizations, particularly women's groups, should be reflected in energy planning. The "people's science" approach – science and technology for and by the people – is relevant here. The role of a catalyst – a person, group or institution who brings the ideas and needs of local people to the attention of planners and policy-makers, and involves them in development – can be crucial. A holistic perspective is required to ensure that energy planning becomes a central element in rural development (Bajracharya, 1983).

CASE STUDIES

AFI: A WOMAN FROM GHANA

Ghana's environment is suffering the effects of dramatic changes: its forests have degraded into savanna, and the savanna areas are fast turning into deserts. The invasion of the desert through overcultivation, forest clearing and overgrazing has been worsened by extreme changes in the climate of West Africa since the recent severe and persistent droughts. Vegetation has become so impoverished that it is difficult for the forests to recuperate even with the onset of the rains. At the current rate of deforestation, about two-thirds of the country will soon experience severe wood shortages and serious environmental resource degradation.

The most affected areas in Ghana are the northern coastal grasslands and the Guinea savanna areas in the south. The situation in the south is further worsened by the high price of fossil fuels which in turn brings greater pressure on woodfuel resources, especially in the southern savanna villages near large urban centres. Deforestation has damaged water resources, brought about erosion and lowered agricultural productivity. Rapid population increase (some 2.4 per cent a year) adds to the pressures.

In the centre of this problem are women. Their domestic and occupational roles are changing not only with the transformations taking place in the development of the country but also as a result of this environmental degradation. Afi, a woman in a savanna grassland village in Ghana, is one of those whose fortunes have changed.

A farmer in Ghana

Afi is a middle-aged woman with two children, a daughter and a son. They live in Ashale Botwe, a settlement in the heart of the plains of the Greater Accra Region. The village is set in a low-lying undulating landscape, and seems isolated although there is a good, easily accessible road to the nearby town, which is the commercial hub of the country. The link with Accra is also political and historical. Afi is one of the indigenous Ga people of the village who constitute a majority of the population, but one of the poorest. She is a farmer in contrast to the most privileged groups represented by the cattle owners, the cattle herders, and the civil servants.

As a farmer, Afi depends entirely on the seasonal rains. Under normal conditions she clears the land of trees between December and February. With the onset of the rains in March and April, the wood is burnt and the land is prepared for the planting of vegetables, mainly tomatoes and okra. Cassava and corn are also sown. The harvesting of the vegetables reaches its peak in

July, after which she sells most of the crops for cash to purchase other food and goods for the family.

Life was rough for Afi during the worst drought in living memory. She lived in an area with limited and unreliable rainfall, and the effects of the drought were devastating. As crops failed and there was a general shortage of food and cash, the provision of daily meals for the family was a major problem. Afi needed other sources of income to buy food that would otherwise have come from her farm. Some women farmers went into commercial fuelwood production but Afi did not find this venture attractive as she would have had to work several weeks before earning anything. To survive, she needed money daily.

Afi's day
So Afi began a small-scale charcoal-production business – a commercial success in the village because there was a shortage of kerosene and bottled gas used by many people in the nearby city of Accra. Her business claims most of the day. She rises at about 5 a.m. and by 6 a.m. is on her way to the savanna woodland to collect fuelwood for the daily load of charcoal. This is the most energy-sapping stage of her industry. As wooded trees are scarce, their constant and continuous exploitation has increased the distances Afi must walk. Whereas formerly she walked less than a kilometre, now she walks three and even more to the production site where she searches for trees with branches large enough to be exploited for charcoal. It takes her 30 minutes to find mature trees. Normally she would have left this land fallow for ten years or more, and the trees would be the future fuelwood for the family. But now all the trees will be gone within a few months. Her only concern is how to survive with her family.

After cutting the wood into small logs she piles them into a huge enamel container and headloads it home. The return trip takes about 45 minutes because of the load. In all, fuel production takes about five hours; she gets home between 11 o'clock and noon. In the production and transport of the fuel she is assisted by her daughter and son, who carry nearly as much together as their mother. As soon as they arrive the charcoal-making begins. Afi sorts and arranges the fuelwood, piling it carefully and covering it with leaves and earth to induce carbonization.

Charcoal-making needs constant attention to prevent the fire burning with a flame and the mound developing a crack. While the process goes on Afi and her daughter do the cooking, which takes about three hours. By 6 o'clock in the evening the family meal is served and Afi goes back to the manufacturing site to check on the charcoal. The process takes about eleven hours excluding the night-watch. In spite of all this labour, her efforts produce only about 10 kilograms of charcoal a day.

Early the next morning she wakes up again at 5 a.m. to sort out and pack the charcoal for sale. She leaves the house at 6 o'clock and makes a five-kilometre trek to the nearest market to sell a small basket of charcoal. After the sale she buys groceries and fish for the daily meal. On her return, she treks again to the savanna woodland where another process of charcoal-making is set into motion. Her laborious task would be even more burdensome without the assistance of her children: her daughter is responsible for most of the house cleaning, water collection and other household chores, thus leaving Afi free to devote more time to her business. The children also help her in the production of fuelwood: she is the first to leave, but later her daughter follows after she has finished with the laundry and the other housework. Sometimes, Afi's husband comes along, when he is not working, to help with the production and to carry a headload.

How Afi copes

The constant exploitation of trees has put so much pressure on the vegetation that Afi's choice of wood species for fuel and charcoal has shrunk. Now she has to choose from only three different species. The Nim (*Azadirachta indica*), a drought-resistant tree, is her favourite source because it is widely available, produces excellent fuel and makes good charcoal. Haatso (*Fagara xanthoxyvloides*), though not as popular as the Nim, is also highly combustible and good for charcoal and so is Nokotso (*Diospyros mispiloformis*). But Afi scarcely uses these two species because they are not available on her fallow land. The increasing scarcity of wood compels her to produce fuelwood only from her farm, where she has cutting rights.

The problem of fuelwood resources encourages Afi, just like other women in the village, to manage the wood carefully. She leaves the farm fallow to rest for a long period of time to regenerate. To assist regrowth and prevent soil erosion after felling, tree stumps of about two feet high are left standing. This is her major method of sustaining the vegetation. But, because of the constant pressure for wood and the need for agricultural, commercial and residential lands, the fallow has been drastically shortened. The result is that the trees are not given enough time to recuperate fully before they are cut. Occasionally she used to plant a shade or fruit tree, but she has not done so for a long time. The number of trees on her farm is dwindling.

Environmental deterioration and human welfare

Low agricultural productivity and crop failures have put Afi under stress. She experiences frequent food shortages and has to put her husband, children and grandchildren first when serving meals. Most of the time she satisfies herself with the barest minimum. The food is low in fat (which is a concentrated source of energy) and protein, for fish, meat and eggs are costly. The deficiencies and the low calorie intake reduce Afi's resistance, affecting her health and that of her family. She complains often of malaria and is generally unwell.

Despite her poor health she has to struggle with the charcoal business to obtain cash to consult a doctor and buy drugs as well as food and other essentials. She is caught in a vicious spiral of poverty. The most hectic time is just before the next planting season; this is also the period when she has to work hard to sustain the family. When the rains fail, her workload increases as other household chores like water collection become a problem. The charcoal business, which is her only hope for survival, causes constant fatigue. Felling trees in the sunshine brings an aching body, headaches and fever, while the burning of charcoal over long periods in the smoking environment makes her sick and gives her backache from the constant stooping. Only conservation and sustainable development offer hope to women like Afi in the rural areas.

Source: Elizabeth Ardayfio-Schandorf, Geography Department, University of Ghana, Legon.

THE LIFE OF IONE HALLEY, GUYANA

Ione Halley, mother of nine children, lives with her husband in the village of Manchester on the Corentyne coast, 85 miles east of the capital city of Georgetown, Guyana.

Manchester is an agricultural village; most of its families cultivate rice or rear livestock. The Halleys depend for their livelihood on pig-rearing and, according to Ione, it pays. But life on the farm is not easy. For a great part of each day, she and other family members (mainly the children) have to be away from home gathering wood for cooking and for the preparation of animal feed. Here, as in other rural areas of Guyana, fuel scarcity is reducing the ability of women like Ione to meet their basic family needs. More than half of Guyana's households use the traditional open fireplace, made from mud, cow dung or cement blocks and fuelled by firewood, coconut husks and charcoal. There have been some improvements with designs adopted from India that incorporate a chimney and reduce smoke, but these fires still require a supply of wood or coal. Tending them can take up to five hours out of each woman's working day. Some families use bottled propane gas to supplement their woodfuel for cooking, but shortages are common and people often have to wait long hours at the distribution centres.

Ione used to use a brick "chula" fireplace, of improved design, fuelled by wood or coconut shells. Every week, the family spent the equivalent of US $4 on extra wood – a sum it could ill afford. The small kitchen was always full of smoke which Ione knew was "bad for the eyes". For Ione and her family the day was a constant battle of carrying water, collecting firewood and dealing with the animal waste in the compound where her 80 pigs were housed. They began work at 5 a.m. and finished at 10 p.m. The nine children helped with preparing the meals, keeping the pigs fed and clean, and collecting the fuel, but that left little time for play. What her family needed, thought Ione, were safer

and easier ways of procuring a reliable source of energy. And it was in the
search for alternatives that she and her family became involved with the United
Nations University and the Institute of Applied Science and Technology, which
began a Biogas Programme in Guyana in 1984. The chance came for Ione to
participate in an experiment which would make use of the animal wastes that
polluted the environment.

The aims of the Biogas Programme were to:

- improve the quality of life by increasing agricultural and industrial
  production;
- conserve the surrounding environment; and
- reduce the time spent in gathering fuel.

In the autumn of 1984, a video about the operation and benefits of biogas was
shown to 450 farmers from the surrounding districts of the village of
Manchester. Three-quarters of them were interested in the possibilities of local
biogas schemes. Farmers were selected for their ability to participate in the
biogas experiment on the following criteria:

- their capacity to finance a digester;
- their access to fresh water;
- their ability to mobilize self-help to excavate a digester pit;
- the potential use of the effluent as fertilizer;
- the number of animals;
- the area of land under cultivation; and
- their cooking gas needs.

Ten farmers were selected for the first part of the programme. At first, Ione
failed to qualify, for her farm had no permanent supply of running water. But
by lobbying hard among the politicians and officials who visited the district,
Ione managed to get water piped to her farm (although she had to buy the
pipelines and supply the labour for ditching). In April 1985, construction work
began on a digester for Ione's farm. Completed, it had cost US $608, with all
the members of her family helping with the building and local villagers
contributing 425 hours of labour.

Now, Ione's family does not need to use the smoky open fireplace. She
bought a new gas stove which was adapted to use biogas; this makes cooking
and baking much easier and gives the family more free time. Ione and her
children built a kitchen garden on a ten-square-yard plot which is fertilized by
the biogas effluent. The produce from the garden saves some of the family
income and adds fresh green vegetables to their diet.

So pleased was Ione with the technology, which she describes as "magic",
that she has opened her kitchen to her wider family and friends, who have been
encouraged to bake on the new stove at weekends. She also encouraged local
schools to participate in the experiment. The children visited her home and

garden to see the digester working, and surplus fertilizer was given to one school for its garden project. All this increased awareness, among children and the community, of the effectiveness of biogas technology. Here was a method of using animal waste to generate gas and fertilizer which saved time and money and ensured a clean environment, both on the farm and in the home.

The lessons learnt were many. As a female farmer, Ione was highly effective in spreading the message to her neighbours, her relatives and young people in the local school. Her example multiplied the effect of the biogas programme. She had demonstrated that political lobbying, persistence and hard work can bring access to needed services like piped water.

The biogas experiment linked many approaches – ideas from China, technical know-how from the Institute of Applied Science and Technology in Guyana and financial assistance from the UN University. But above all, success was due to the commitment and participation of the rural farmers. Ione especially was a key to the effective transfer of a new technology for sustainable development which increased her own agricultural production, and improved the environment in and outside her home.

Source: Sybil Patterson, Women's Revolutionary Socialist Movement, Guyana.

THE MOMBAMBA WOMEN'S GROUP, KENYA
Kiamokana, in Kisii, is a medium-sized town in the second most densely populated area of Kenya. There is an acute shortage of firewood: 95 per cent of the people use wood or charcoal to meet household energy needs. Prices have gone up and a fifth of the household income now has to be spent on fuel alone. People can only afford meals with a short cooking time and traditionally nutritious foods such as beans and maize are now rarely offered.

In 1983, a Special Energy Programme began in Kiamokana with the existing Mombamba Women's Self-Help Group, which had 56 members, mostly women. Later, some husbands joined. The original purpose of this close-knit group was for members to help each other weed and harvest tea and pyrethrum.

Early on, a survey was conducted to assess fuelwood consumption and, importantly, determine the socio-economic characteristics of the community. Lasting three weeks, it revealed that, on average, a woman has eight children and most households have more than one wife. Only a few children go to school; girls marry young and start their own families. Most people are poor, though there are some middle-income and rich families. While the rich use kerosene for lighting and charcoal for heating, the majority depend for their fuel upon agricultural waste such as twigs from the tea and pyrethrum bushes. The need for a project to improve cooking efficiency was obvious.

In 1984, the Maendeleo Ya Wanawake Women and Energy Project was launched with funding from the German Technical Agency GTZ to promote improved stoves. Salima and Mary, two members of the Mombamba group were trained to construct the stoves, which are built around clay liners and cost about 25 to 30 Kenyan shillings, depending on the design and size. Mombamba members who wish to obtain a stove approach the group. The group leader pays the money to the project officials, who release a liner of the required size and shape; a trained worker, with the help of the community, builds the stove.

So far, more than 1,000 one-pot and two-pot wood-burning stoves have been built in the area. Salima and Mary and other extension workers receive a monthly payment as an incentive to carry out training and other activities. Over 100 women have been trained to construct the new stoves, and neighbouring villages are now also involved. Their popularity is understandable: not only do the new stoves consume less fuelwood more efficiently than the older stoves, they result in a cleaner kitchen and a healthier environment.

With the help of the Kenyan Ministry of Energy's Agroforestry Centres, Salima, Mary and three other women have received training in agroforestry techniques and have established nurseries and firewood plantations. Salima has distributed 500 seedlings to members of the Mombamba group and saved 1,000 for a healthy plantation on her own plot, which produces seedlings for sale. Most of the people like to grow crops between their trees which they can harvest for fencing.

Now, more women are learning agroforestry techniques; many tree nurseries, which generate income and help to arrest soil erosion, have been established. Trees, especially those suitable for firewood, are being planted in backyards and communal lots all around the village. Another successful income-generating activity of the group is the operation of a posho (grain) mill. Future plans include establishing health clinics and procuring piped water from a nearby tea factory.

Community participation has been the key to success on this project, which has combined the conservation of energy with improvement of the environment, economic use of meagre resources and the development of new income-generating activities.

Source: Prahba Bhardwaj, from information given by Juliet N. Makokha, Energy Programme Officer of Maendeleo Ya Wanawake.

CHAPTER SIX

Human Settlements: Women's Environment of Poverty

By the turn of the century, almost half the world will live in urban areas – from small towns to huge megacities (WCED, 1987).

Since the beginning of this century, the world's population has grown from 1,600 million to more than 5,000 million. By the year 2000, the UN predicts a global population of 6,000 million. Growth rates are especially high in the Third World: here populations have been expanding more than twice as fast as those in developed countries.

POVERTY AND URBANIZATION

Most of the world's people still live in rural areas, and the majority are poor. FAO estimates that 70 per cent of rural people in the Third World (excluding China) live on small farms or are landless (World Resources, 1986). With some 1,400 million people living in abject poverty, more and more are migrating to urban areas, attracted by the prospects of employment and a better standard of living. Others leave their villages to join a relative, find a spouse or escape from a difficult family situation: for young people especially, as the "Child from Delhi" case study illustrates, the city seems to offer freedom.

Rapid urbanization is not caused only by migration from the countryside: Third World cities have high rates of natural population increase. At the start of this century, less than 14 per cent of the world's people were urban dwellers; by 1985, the proportion was 42 per cent, with over half living in the cities of the Third World. The largest agglomerations have shown a spectacular expansion. Since 1950, the populations of cities such as Abidjan, Dar es Salaam and Nairobi have increased sevenfold. It is estimated that by the end of the century some 439 cities will have more than a million people, and nearly 100 may have more than four million each (UN, 1985; UNFPA, 1986). Beijing, Buenos Aires, Cairo, Mexico City, São Paulo and Shanghai already have populations over 10 million. Although some commentators advise caution in projecting the future patterns of urbanization in the Third World, especially for large cities, the trend is clear: increasing numbers of poor people, coupled with diminishing resources to cope with their needs and expectations (Hardoy and Satterthwaite, 1986-1).

Some countries are now promoting the development of small and intermediate centres in their settlement strategies, but this offers little hope of alleviating the appalling and deteriorating conditions in most Third World cities.

Environments of poverty

Half the inhabitants of some cities in the South live on vacant lots in the urban centre or as squatters in the slums and "popular settlements" at their margins. Here, people provide their own shelter, often illegally, and frequently on land that is unfit for housing and prone to severe environmental damage. To avoid eviction, the poor build on land of low commercial value. Homes are constructed upon unstable hillsides or on land subject to flooding by sea and river. In Mexico City, about 1.5 million people live on a dry lake bed which floods in winter and creates dust storms in summer. Houses may be located on sandy desert land or in heavily polluted areas close to industry.

In such overcrowded, makeshift settlements, there are few, if any, services save those organized by the slum-dwellers themselves. Often there is no water supply, sewerage, garbage removal or electricity; roads and transport facilities are poor, medical care is inadequate and there are few schools. Debt and the other economic crises faced by Third World countries ensure that there is little or no investment in construction, basic infrastructure or social services.

The World Commission on Environment and Development describes conditions in India:

> Out of India's 3,119 towns and cities, only 209 had partial and only 8 had full sewage treatment facilities. On the river Ganges, 114 cities each with 50,000 or more inhabitants dump untreated sewage into the river every day. DDT factories, tanneries, paper and pulp mills, petrochemical and fertilizer complexes, rubber factories, and a host of others use the river to get rid of their wastes. The Hoogly estuary (near Calcutta) is choked with untreated industrial wastes from more than 150 major factories around Calcutta. 60% of the city's population suffer from pneumonia, bronchitis and other respiratory diseases related to air pollution (WCED, 1987).

Those who live in these "environments of poverty" (Hardoy and Satterthwaite, 1984) suffer three kinds of environmental degradation, the constant danger of industrial pollution, the effects of minimal basic services and the cumulative deterioration of the urban hinterland.

Industrial pollution

Many popular settlements are located in environmentally vulnerable or dangerous places, thus increasing the chance of catastrophe. And, as the industrial accidents of the last decade show, the danger is ever-present and profound. In 1976, dioxin released in Italy's Seveso disaster poisoned thousands. Mexico City's poor San Juanico neighbourhood was devastated by an explosion of liquefied petroleum gas in 1984: 452 were killed, 4,240 were injured and more than 31,000 were made homeless. At least 2,500 people were killed by the methyl isocyanate gas which leaked from a pesticides plant in Bhopal, India, in 1984. The vapour which escaped into Bhopal's slums has left many others permanently damaged. A Third World nuclear accident on the scale of the Chernobyl explosion in May 1986 would bring much greater devastation.

Many industrial plants in the South fail to meet the strict safety standards required in the North. Most are located in dense population centres (UNEP, 1985). Not only is there a continual risk of explosions, but industries are also a permanent source of pollution in these areas, for the rapid industrialization of the South is marked by the transfer of "dirty industries" from the North. Air pollution from noxious emissions and water pollution caused by waste dumping in local waters have serious health implications, both directly within neighbourhoods and farther away, beyond city boundaries. And there are other consequences: in India, Malaysia and elsewhere, river and coastal fisheries (and therefore livelihoods) are destroyed by industrial effluents. The soils surrounding industrial plants, on which slums are built and food may be grown, often contain high concentrations of heavy metals.

Poor services

The lack of clean water, sanitary systems and solid waste removal means that most informal settlements have no option but to pollute themselves. Millions literally live on city garbage dumps. Their health is directly affected: intestinal diseases flourish and are rapidly transmitted in the cramped living conditions. The high levels of air pollution from traffic (especially from old and badly maintained engines), as well as industry, cause serious respiratory problems. In Jakarta and Mexico City, for example, respiratory ailments are legion.

Human settlements have a major impact upon the environment of their hinterlands: they can destroy the soils, crop and forest lands which support life, provide energy and sustain incomes in the informal economy. Around the cities of the South, as explained in Chapters 4 and 5, deforestation, overcropping and overgrazing often result in permanent degradation of the soil, which is followed by erosion, flooding or the encroachment of desert.

Refugee camps

Some of the worst environmental effects of human settlements are found at the margins of refugee camps, where people who may already be victims of natural disasters – such as the Sahelian drought – are desperately trying to meet their basic needs. Around the camps, demands for fuel as well as increased "slash and burn" practices to grow subsistence crops can bring total environmental degradation. In camps in Somalia, fuel became more valuable than food, and a "fuel-for-work" programme (as opposed to food-for-work) had to be introduced to allow the forest to recover (Cecelski, 1985).

Where refugees are moved to new regions or countries, the clash over scarce resources can be profound:

> In Pakistan, nearly three million Afghan refugees have settled, many with their animals, in the already semi-desertic North West Frontier and Baluchistan provinces. Bare hills now surround most of the camps, and conflicts between the armed Afghans and local Pakistani villagers over forest resources have been a constant point of friction (Cecelski, 1985).

EFFECTS ON WOMEN

> We sit here day after day worrying what will happen. I wonder if the rest of my life is going to be like this. We don't have a real life here, we just survive. I would like to go home, where we can grow our own food (Sara, an Ethiopian refugee of 14 living in a Somalian camp).

Women make up a slight majority of the global population and a visible majority of the poor. A large proportion of the world's homeless – about 1,000 million people – are women (Celik, 1985). And of the world's ten million refugees, women and children make up, in some areas, 90 per cent.

As home-makers and child-rearers, women are directly affected by the place in which they live (Kudat, 1986). The prospects for their children are slender: poor people in parts of many cities can expect to see one in four of their children die of malnutrition before they are five. Just as the lack of sanitation in the home affects women most strongly, so the destruction of water resources and fuelwood around settlements hits them hardest. The environment, both inside and outside the home, has everywhere a greater significance for women – a fact which is reinforced by cultural patterns.

In some areas it is still a woman's task to supply the building materials. In Ghana, for example, rural women haul mud for walls and thatch for the roof, and often carry out the final plastering (UNCHS, 1985; Agarwal, 1983). But there have been few studies of the nature and size of the burden women face in providing shelter. Clearly, some groups are especially disadvantaged, including the old and the growing number of women heads of household. Women now make up 90 per cent of single parents.

These groups are commonly excluded from shelter programmes, for women suffer here, as on other issues, from a general discrimination. With little or no formal education and training, it is harder for them to be employed outside the home. When they are, their incomes are low. As elsewhere, their access to land and credit is limited. Although traditionally, in many cultures, women have been the main builders and maintainers of the family home, men dominate the modern approach to shelter planning and construction.

Inside the home

Women spend more time at home than men, much of it in maintenance and repair. Their daily domestic tasks – cooking, washing and child care – confine them to the environs of the house. Women in seclusion, particularly in Islamic countries, must remain inside most of the time. Yet the internal environment of the home is often dangerous: fumes from open wood and charcoal fires damage eyes and lungs. Household accidents are common where there is little protection from unguarded stoves and heaters, and where perhaps seven or more live in one room (see Chapter 5 and the case study on women in seclusion).

Environmental disasters, too, can affect women more severely, although their perspective is rarely studied. In the aftermath of Bhopal, many suffered menstrual and other gynaecological problems; pregnant women gave birth to deformed babies; some of the infants were blind. In the demonstrations which followed the gas leak, women were especially active in protesting against the damage to their families (Butalia, 1985).

Women at work

Often the workplace is no better than the home. Increasingly, multinational firms are employing young women in the South to work in electronics and textile factories. These women have few legal rights – and even fewer rights in practice – and have little power with which to fight workplace hazards. Often they work with toxic chemicals, or damage their eyesight by continuous use of microscopes. Lighting and ventilation are inadequate and there is usually no protection from machinery, dust and noise. The hours are long; rarely is there a right to refuse overtime. Where employers provide living quarters, as in many Asian countries, women live in cramped company dormitories, sharing beds in tiny rooms. Worldwide, women workers are among the lowest paid and least organized in unions, which makes them vulnerable to these kinds of exploitation and unable to secure better living and working conditions for themselves.

RESPONSES TO THE PROBLEMS

Shelter is now officially recognized as a basic need. More and more local housing organizations are encouraging community participation and are aware of the need for legal access to land, building materials, credit without collateral and incomes to sustain investment in better housing (Hardoy and Satterthwaite, 1986; Lusaka case study). The inhabitants of many informal settlements display extraordinary vitality and resourcefulness in providing shelter and services for themselves and in generating incomes. In the poor districts of many cities, communities are transforming their environment through self-help.

But they cannot continue to do so without supportive policies and legislation. Outdated by-laws prevent the poor from getting access to affordable shelter and from earning a living in the informal economy (in some cities, petty traders are hounded like criminals). Strict application of outmoded regulations limits the adoption of more appropriate technologies. For example, costly sewers instead of latrines are installed in areas where there is no water for flushing. Roads are built but then not maintained. Jorge Hardoy of the International Institute for

Environment and Development argues that most present shelter priorities in the South are misguided: money spent on expensive housing and service schemes should go instead to support comprehensive and affordable self-help shelter schemes organized by NGOs. Governments and aid agencies need to assist the poorest by granting them legal tenure of plots to build their homes, by providing building materials (and promoting their local production) and by increasing access to credit. "People," says Hardoy, "are themselves a resource."

"Site and service" and settlement upgrading projects (as described in the Lusaka case study) adopt this philosophy. Even here, however, the needs of women can be ignored. Moser (1985) observes that housing projects often use a gridiron layout that does not allow women to work in their house and at the same time keep an eye on their own or their neighbours' children. House designs and plot sizes rarely consider the fact that many women will want to use their houses as workshops or as shops to sell goods – such enterprise is often forbidden in low-income housing projects. Hosken (1987) argues that housing managers as well as international financiers must start to recognize women's vital contribution in building new communities and maximize the opportunities for them to be full partners at every stage of the work.

UN Decade for Women
In 1975, at the start of the UN Decade for Women, the Plan of Action of the Mexican Conference referred to the special role of women in human settlements. Official recognition came at the 1976 UN Habitat Conference in Vancouver. Ten years on, at the Nairobi Conference which marked the end of the UN Decade for Women, documents were still stressing the need for women to participate in and benefit from plans, programmes and projects related to the development of human settlements. The resolutions adopted in Nairobi, published as *Forward-Looking Strategies for the Advancement of Women*, called for a number of major changes of policy and practice:

- The enrolment of women in architectural, engineering and related fields should be encouraged, and qualified women graduates in these fields should be assigned to professional, policy-making and decision-making positions. The shelter and infrastructural needs of women should be assessed and specifically incorporated in housing, community development, and slum and squatter projects (paragraph 209).

- Women and women's groups should be participants in and equal beneficiaries of housing and infrastructure projects. They should be consulted on the choice of design and technology of construction and should be involved in the management and maintenance of the facilities. To this end, women should be provided with construction, maintenance and management skills and should be participants in related training and educational programmes. Special attention must be given to the provision of adequate water to all communities, in consultation with women (paragraph 210).
- Housing credit schemes should be reviewed and women's direct access to housing construction and improvement credits secured. In this connection, programmes aimed at increasing the possibilities of sources of income for women should be promoted and existing legislation or administrative practices endangering women's ownership and tenancy rights should be revoked (paragraph 211).
- Special attention should be paid to women who are the sole supporters of their families. Low-cost housing and facilities should be designed for such women (paragraph 212).

Hosken further defines poor women's requirements in low-cost settlement projects. Women and men should be guaranteed equal access to plot ownership or leasehold. There should be no plot use restrictions on running small businesses from the home (unless that interferes with a neighbour's rights or creates a public nuisance). Financial support must be provided so that women can develop individual enterprises or cooperatives. Plot sizes should be large enough to grow vegetables and/or keep small animals for food, or communal garden plots should be included in the settlement plan. And women should be fully represented in community organizations which are concerned with management (Hosken, 1987). In these ways, women can be enabled to play their part in building more productive and self-reliant settlements.

At the NGO Forum in Nairobi, workshop discussions on Women and Habitat led to the formation of an international network on this issue; there has been some follow-up in seminars and workshops. The UN International Year of Shelter for the Homeless in 1987 should have increased awareness. But action still lags far behind the talking: in most cities of the South, the position of women has worsened. At the same time, however, local grassroots initiatives show that change is possible.

Local action

In Indonesia, the traditional Arisan self-help system among rural women combines basic housing, financed by government, with improvements (bathrooms and latrines) financed by the women themselves. In Jamaica, few projects are designed specifically to help women, but there is one exception: the Women's Construction Collective, which has trained young unemployed women in building skills and found them jobs (see case study). In a similar project in Panama, launched with support from the Ministry of Housing in 1981, poor women (half of them single parents) aim to construct 100 homes. They receive land, building materials, training and a small monthly wage. By minimizing labour costs and bringing new skills to the participants, the scheme appears to have served both the government and the women well (UNCHS, 1986).

When the chance is offered, it is clear that women are anxious to respond. In two low-cost housing projects in Zimbabwe, the majority of participants in the workshops on community issues were women. They now play an important role in the design and management of their houses and in the shape of the overall settlement programmes (Report, 1980). In Kenya, the Mabate Women's Groups are among a number of active self-help building cooperatives, and the Kenyan Women's Finance Trust intends to start up a loan-guarantee scheme to help women finance shelter improvements.

Nor is construction the only sector in which women can influence their living environment. In Mexico, they played a crucial part in adopting and promoting an integrated system for recycling wastes (SIRDO), a scheme that now controls sanitary conditions and offers the local community the possibility of earning an income. Pakistan's Baldia project (see case study) has transformed the prospects for women. And, as Chapter 3 discussed, they are involved in the provision of clean water.

Urban agriculture

While women remain largely excluded from the benefits of the formal economy of cities, their position in the informal sector is improving. They play a dominant role, for example, in selling "street foods" (Tinker and Cohen, 1985). In Senegal, women constitute some 53 per cent of vendors, although both the food and the place of sale are completely sex-segregated. In the Philippines, women control 79 per cent of street food enterprises and, in the seven per cent that are owned by couples, are the major decision-makers. With as much as 20 per cent of the food budget being spent on street foods, according to a worldwide study conducted by the Equity Policy Center, these small business ventures can be an important source of income for women. By creating demand for locally-grown produce, they also provide income to rural farmers, particularly to women farmers, with whom women street vendors often collaborate (Irene Tinker, EPOC).

Where low-income families can cultivate a vegetable garden, the crops can contribute substantially to the family diet and supplement the family income. As the Lusaka case study shows, urban gardens are a priority for many women, especially when incomes are low. Urban agriculture is beneficial on several counts: it provides fresher, cheaper food, more green space, and scope for recycling household wastes, all attributes highly valued by women as the providers of food and the ensurers of good health in the family.

Much more vacant urban land (of which there are large amounts in some Third World cities, especially in Africa) could be cultivated if more women were granted legal title or access to sites and supported in other ways, for example with seed, tools, water and fertilizers, storage space, advice and credit for improved production and marketing. Demonstration plots and training schemes are needed, but few settlement projects have the staff, funds or institutional capability to incorporate such developments (Davidson, 1985).

In many ways, urbanization offers a unique opportunity for women to change their lives and escape from some of the oppressive traditions of the past and of village life which have excluded them from positions of control and stifled their initiative (Hosken, 1987). Living in the city offers women the prospect of education and self-development, of learning new skills, of earning an income. Information about health care, family planning and women's rights is more likely to be available. But, as a priority, women's access to all these aspects of urban life must be improved.

It is clear from the success of some local projects that community-based organizations and NGOs do well when they are helped and encouraged to involve women in the provision of shelter, and in the development activities that flow from it. The World Commission on Environment and Development argues for a much larger proportion of development assistance to be channelled through these organizations (WCED, 1987).

CASE STUDIES

A CHILD OF DELHI

Urmila is 14 years old. Five years ago, she moved with her family from a poor village in Uttar Pradesh to India's capital. Now she is a child of the city, and very unwilling to return to her village.

"I don't want to go back," she says. "There is more entertainment here – TV, video – and we have better food."

Not that families like Urmila's live in very attractive conditions. Old Delhi is a labyrinth of small streets studded with mosques, temples, monuments, bazaars and wandering cows. There are about 4,000 people per square kilometre, and 40 per cent of the population live in slums – the "bastis".

Urmila's family came to Delhi because her father was sick with asthma and it was thought he would get better medical treatment in the city. Urmila, who was nine when they arrived, her 13-year-old sister and her 10-year-old brother immediately looked for work. They spent every day running errands, washing utensils or carrying goods through the narrow streets. Their day began about 6 o'clock in the morning when their mother started to prepare the day's meal. By 8 a.m. they would all have left for work, except the father, who was ill and stayed home to look after the baby.

At 6.30 in the evening they returned. Then the hut was lit by a paraffin lamp and food was cooked on a charcoal stove (the girls collected unburnt coal from the railway station nearby). By 9.30 p.m. they were all asleep.

After two years the routine was broken by the father's death and the family split up. Urmila's elder sister returned to the village to marry; the baby was sent to a relative; Urmila, her mother and younger brother all went to work in Delhi as domestic servants. They now live with their employers and rarely see each other. Urmila earns Rs 150 (about US $20) a month in the home of a wealthy industrialist. She looks after the young children, plays with the older ones, cleans, sweeps, dusts and cooks. Her day begins at 6.30 a.m. and ends at 10 p.m. But Urmila feels she is much better off – and healthier – than she was in the village, although she would like to see her family. The only thing she does not like about the city is the traffic.

Urmila has less in common now with her mother, for she has been learning the language of her employers while her mother still speaks the village dialect. But Urmila knows that she may well have to return to the village to marry. "If my mother forces me to, I'll have to go, there would be no choice."

Source: Atiya Singh, 1986, UN Fund for Population Activities.

THE WOMEN OF BHOPAL

On 3 December 1984 a lethal gas escaped from the Union Carbide pesticide plant in Bhopal, India, killing more than 2,500 people and severely injuring another 20,000. Women continue to suffer the ill effects of the gas leak. High rates of spontaneous abortion and the birth of deformed babies were a direct result and the women of Bhopal continue to be plagued with uncertainties about pregnancies. But it seemed that nobody was really concerned about the ill effects of the gas on the women themselves.

Two women doctors carried out a clinic-based survey in March 1985. They showed that since the gas exposure, an extremely high proportion of women had developed gynaecological diseases, such as leucorrhoea (94 per cent), pelvic inflammatory diseases (79 per cent), excessive bleeding (46 per cent), and lactation suppression. But the medical establishment ignored this study. In response, a progressive medical group carried out a well designed survey on the health problems of the affected women in September 1985. The survey showed clearly that the rate of spontaneous abortion in women who conceived after the gas leak was significantly greater than the abortion rates prior to the leak. Alterations in menstrual flows also continued to be significantly high. But these gynaecological disorders were not recognized as such by the medical profession in Bhopal and were therefore untreated.

Socially, in many instances, women are also being discriminated against. Women have been sent away from their husbands' homes as the gas had affected their ability to work; 65 per cent of working people in the slums of Bhopal experienced a drop in income ranging from 20 per cent to 100 per cent. Women have been divorced because of the fear of conceiving abnormal babies, or because they can no longer have children.

Source: Dr Sathyamala, "Lest we forget the Women of Bhopal", in *Women's World* (ISIS), no. 14, June 1987.

INVISIBLE WOMEN: PURDAH IN PAKISTAN

Purdah is the Persian word for curtain. It is an institutionalized system of seclusion and veiling which operates at three levels: the physical segregation of women's living space which is the secluded world of the zenana (women's quarters); social segregation, which allows women interaction only with the immediate kinship circle; and the covering of the female face and body.

The purdah system keeps women secluded and isolated from the wider world. Their environment is that of their home, family, immediate kinship circle and neighbourhood. In urban centres women are segregated at work, travel to and from work in separate sections on public transport and are covered in purdah clothing at all times. There are separate wards in hospitals, and schools and colleges are segregated, as are religious and marriage ceremonies. Among the urban middle class, purdah is a status symbol for those who can afford to keep their women in seclusion.

Rural women who work in the fields also observe the purdah system, although there are fewer restrictions on their movements. One reason for less strict purdah in the rural areas is the lack of differentiation between the home and work place. Women are involved in the production of crops and their processing before the produce is marketed, so the question of going out to work does not arise: women work at home. Even so, women remain clad in heavy chaddars despite the blazing heat and exhausting work. And the ideology of the purdah system remains: social interaction, for example, is still restricted to the family.

Purdah is imposed at puberty and from then on the private and public spheres are clearly defined. The respect (izzat) due to a family depends on the behaviour of its women. To deviate from the prescriptions of purdah is to dishonour and shame the entire clan. In conventional homes, separate rooms for women form the zenana, to which unrelated males have no access and which related males do not enter unannounced. The zenana lies at the back of the house behind the men's quarters (mardana). The private sphere thus symbolically represents a patriarchal social unit where the man as "protector" occupies the space in front of the women. These "separate worlds" dictate the division of labour in the family and innate differences are presumed between the protector and protected, from which an assumption of fundamental sexual inequality is generated. In families where physical segregation does not exist, purdah manifests itself in clothing, gestures and the silent movement of women (Papanek, 1982).

The most powerful image of purdah is women's clothing. In Pakistan, the "burga", which consists of material attached to a skullcap, covers the wearer from head to foot, the only opening being a netted section for the eyes. The "chaddar", an ankle-length shawl, also covers the wearer completely, though the face or eyes can be left exposed. The "dupatta", a two-metre-square scarf, drapes over the shoulders and usually covers the head. The segregated world of women in both urban and rural areas leads to ignorance of the outside world, illiteracy, loneliness, and a fear of moving beyond the confines of home. Women in purdah may never have learned to negotiate road traffic or go to a doctor alone. Sometimes they remain unfamiliar with the streets of their own neighbourhood.

Women's education, employment and health are all affected by purdah. Girls have to leave school at an early age, or are not sent at all if there is no segregated school. Less than 16 per cent of urban women and only 6 per cent of rural women are literate in Pakistan. Provision for more girls' schools with women teachers is being made, but while improving literacy, this also reinforces segregation.

Surveys show that in Pakistan in 1975, 58 per cent of employed urban women and 40 per cent of employed rural women were working as weavers and tailors, with over 95 per cent in both sectors working at home in seclusion. There are 3,000 industrial centres which employ women in handicrafts and embroidery, but the skills they learn, such as sewing and cutting, cooking and gardening, are all extensions of their familial role.

These "industrialized" homes are an example of the complexities of purdah society. They provide women with employment and are the initial step in their recruitment into the labour force, but at the same time they isolate women and reinforce their segregation. Yet, given the rigidity of purdah and its strong traditions, there is no other way of penetrating the home and employing women.

Purdah imposes severe strains on women's health: tuberculosis and other respiratory problems can be aggravated by covering the nose and mouth for long periods. The narrow, netted section of the burga means women have to breathe through folds of cloth whenever they go out and are not able to breathe fresh air freely. Women often carry their babies under their burgas which means the child, hidden under yards of cloth, has to breathe its mother's infected breath. Vitamin D, of primary importance for bone formation, is naturally assimilated through exposure to the sun. In spite of living in climates where the sun shines almost every day, women in purdah are deficient in vitamin D and rickets is common. In addition, most burgas are black, a colour that rapidly absorbs heat. The burden is obvious for women in the hot, humid countries of the Middle East, South Asia and South-East Asia.

A further health problem is the rule that women must visit only female doctors. When there is no female doctor, the woman must suffer in silence. In extreme cases, women can communicate with a male doctor through a male relative who acts as an intermediary, or she can speak directly to the doctor from behind her veil or a screen. But this interaction is confined to a dialogue and does not permit any physical examination.

Mental health is also threatened by the purdah system. Since women's activities in the public sphere are so restricted, they remain inactive, isolated from society and current events. They become lonely and frustrated; their creativity is destroyed and their growth and development as human beings is stunted.

No other social tradition so strongly defines and restricts women's environment. While purdah is being debated in the Muslim world, the ideology remains deeply rooted in the culture and is an integral part of the Muslim identity. Any development programme which hopes to succeed, therefore, must take into account the values on which purdah is based. Social change is a slow process, and the purdah system is a powerful force with which to contend.

Source: Meher K. Marker, IUCN Pakistan.

THE IMPACT OF SQUATTER UPGRADING ON WOMEN:
A CASE STUDY FROM LUSAKA, ZAMBIA

In 1963, the year before Zambian independence, the capital city of Lusaka had a population of 123,000. In colonial times, migration to urban areas was restricted to men with urban jobs, but in the years immediately after independence, the restrictions were lifted. Jobs in administration, manufacturing and other sectors were available in the towns, especially the new capital, where the rate of population growth became one of the highest in Africa. By 1974, its population was around 401,000, and the ratio of women to men had increased as wives joined their husbands. But the official low-income housing programmes had failed to keep pace with the growth and 42 per cent of the city's population had housed themselves illegally, on public and private land around Lusaka.

Women in the squatter areas

Many of the squatter settlements were reached by tracks, with only footpaths to the houses. These were built mainly of sun-dried brick with corrugated-iron roofs. Water came from shallow wells and pit latrines were used for sanitation. Houses were built incrementally, using family finance and sometimes family labour; some were extended to accommodate tenants. Although most of the men were waged, there were few jobs for women. Those who could, sold vegetables outside their homes or in the squatter markets; others did domestic work, dressmaking, typing and beer brewing or became prostitutes.

Women's lives revolved round their neighbourhood, with occasional visits to the city centre for shopping, to the hospital or clinic, or to relatives living elsewhere in the city or in the village of their birth. When possible, they enrolled their children in primary schools, invariably some distance away, but many children could not get school places and few were able to continue to secondary school. With most of the women, many children and some of the men around all day, the old settlements were, and still are, busy places, for much of the living takes place outside – domestic work as well as social activity. Women kept the space outside their houses swept clean, while the rest of the area between the houses in these relatively low-density areas was used for play, vegetable gardens and fruit trees. Despite the poor access, the distance to bus services and social facilities, the wet and muddy conditions in the rainy season and dust in the dry, the shared and smelly, fly-infested pit latrines, and the danger of mud brick houses collapsing in the rainy season, these squatter settlements enabled thousands of poor families to meet their basic need for housing at low cost and in areas which provided the social support and economic opportunities necessary for survival.

Upgrading the squatter areas

Following demands from the squatters, the government agreed to upgrade their housing and to provide basic social services throughout the settlement. Incorporated into Zambia's Second National Development Plan in 1972, the project lasted from 1973 to 1981, and became internationally recognized as a model settlement scheme.

With the aid of a World Bank loan, basic physical improvements were made in three of the largest squatter settlements, where over 60 per cent of Lusaka's squatters lived. About 26,000 households obtained access to piped water (one standpipe for every 25 houses). Refuse collection began, although coverage and regularity were far from perfect. Roads were improved, bringing buses and closer to the residents, although the vehicles also increased the danger to children and presented a maintenance problem to the underfunded and understaffed engineer's department. Storm water drainage improved conditions in the rainy season and prevented flooding, and street lighting provided better security at night. And there were new social facilities, including primary schools, clinics and community centres. Markets were improved. Squatters' houses were given permanent numbers, ensuring security of tenure and eligibility for 30-year occupancy licences.

Participation

Public meetings were held in all the areas to be upgraded to inform people about the programme and get their support. No attempt was made to discuss the proposals specifically with women; although many attended the meetings, few asked questions or made contributions. The project based its infrastructure programme on the demands of residents expressed politically and through surveys. So many decisions had to be made before the World Bank loan that there was little scope for participation once the work began, although local political leaders had some say in deciding the routes for roads and the location of social facilities.

Overspill areas

Because of the massive construction and redirection of roads, some 8,000 households chose to take up plots in adjacent overspill areas. These offered the same services as in the upgraded areas, but were laid out in regular blocks of 25 houses around a standpipe, with road access to each block. Loans were provided with which the new residents could build two-room houses in a mixture of traditional and manufactured materials. However, the regular layout of these areas made social contact between women more difficult. Most resettled families built a concrete block house. This, because of the high price of materials used and because the loan could not be used for paying a hired bricklayer, cost more than the loan amount; poorer families and those headed by women were under-represented in the overspill areas compared to the upgraded areas.

House building

Once resettled, nearly all families hired labour to build their house, but also worked themselves. One estimate suggests that, on average, household heads contributed 10 days of labour, their wives 14 days and other members of the household seven, during the four to six weeks typically taken to construct a two- to three-room house. A time budget survey showed that before resettlement, men spent an average of 44 hours a week working and women spent an average of 31 hours on housework. After resettlement, both spent less time on social activities and resting and women spent less time on house building and more on obtaining building materials. The first week or two after resettlement was used for clearing and levelling the plot and digging the pit latrine and foundation trenches. This initial unskilled work was fairly evenly shared between men and women (about 50 hours each in the first week after resettlement), but as construction progressed and the work became more skilled, the women played a less important part than the men, with the exception of queuing for building materials – in one week in the survey area this took 54 hours, compared with 29 hours in construction.

Men talk, women work

Most of the infrastructure was installed by large contractors, although the community did some of the less skilled work such as trench digging. It was noticed that, although men attended the meetings and spoke for the community, much of the work was done by women. The savings in payments to the contractor were made available to the residents concerned for construction of further local facilities.

Residents became liable for a service charge of 1 kwacha a month initially, rising to 3 kwacha once the infrastructure was operating, with loan repayments at seven per cent over 15 years. Although these were small amounts, the low incomes and their failure to keep pace with inflation during the 1970s meant that expenditure on other basic needs, especially food, would have had to be sacrificed to make the payments. In practice, there was widespread and persistent default.

Living conditions had undoubtedly improved – clean water was especially appreciated. For the first time, families had security of tenure – a precondition for a long-term process of house improvement. Families who had moved to the overspill areas valued the opportunity to build a permanent house on a larger plot. The greatest danger – of displacement of existing residents by higher-income families, and of the development of sub-letting by large-scale landlords – had been avoided. But even in this progressive, highly successful settlement programme, women's needs, while not precisely ignored, were not taken into account, and the extra burdens of construction and maintenance were assumed largely by women.

Source: Carole Rakodi, University of Wales, Institute of Science and Technology.

WOMEN'S CONSTRUCTION COLLECTIVE, JAMAICA

The other Jamaica
The high-rise buildings and luxury hotels of Kingston are often seen in travel brochures, but the features which are not tourist attractions are its squatter settlements and urban ghettos. In these areas live three-quarters of the urban population, tightly packed into only a third of the total residential area. High unemployment, intense political rivalries, poor living standards, and the consequent social, economic and political problems make up the texture of everyday life.

Women head over a third of Jamaican households – almost a half in urban areas. Yet women's unemployment is double that of men. Unemployed teenage girls commonly become mothers: it gives them adult status. But because they have no marketable skills, they are ill-equipped to cope with the lifelong economic responsibility which parenthood inevitably brings.

Women's Construction Collective
Ten women from Tivoli Gardens, an inner city area of Kingston, Jamaica, came together in October 1983 to start the Women's Construction Collective. The Jamaican construction industry was booming and building trade workers were in great demand. Three-quarters of the young women in the area were out of work and some contractors were willing to employ female labour. But the women were untrained, for, following a change in government policy, they had recently been excluded from training programmes for the building trades.

The Jamaican Working Group on Women, Low-Income Households and Urban Services in Latin America and the Caribbean decided to sponsor the Women's Construction Collective (WCC) to address these needs and provide a mutual support group. The collective's coordinator, Ruth McLeod, became director of the Construction Resource and Development Centre (a non-governmental organization) in 1984. The centre took WCC under its wing during the early days, and organized training. This included a five-week basic masonry and carpentry course using tools provided through a revolving loan fund.

Breaking into a male business
In the building industry, trade work was customarily done by subcontracted trades gangs, hired according to the task to be performed – carpentry, plumbing or electrical work. Men were recruited through their friends,

relations or work mates from previous jobs, while jobs on building sites were reserved for those sharing the same political affiliations as the locality. Members of other parties entered at their peril, unless their skill was scarce or they had strong links with the contractor.

The Women's Construction Collective managed to secure job placements for their first batch of trainees on a government project but were dismayed to hear that the project was cancelled. Anxious to sustain morale and to prove themselves, WCC developed a strategy of "job auditions", offering to work on site without pay for a trial period. Two women were "hired" and after a week had permanent, paid jobs. Other contractors began employing women as unskilled labourers, but soon saw that they were trained, and promoted them.

Benefits of the collective
Contractors found that having women on site reduces violence and increases productivity. They now employ more women. And the collective can move women across political boundaries with no serious problems. WCC has demonstrated that mutual support among local women can create paid work in a male-dominated industry, and enable them to penetrate rigid political and social divisions. Women's confidence, ability and income have grown. WCC has changed the perception of women's role and capabilities.

Rising demands for WCC services generated their expansion into other localities. To avoid identification with any particular political party, they chose two areas, each with a different affiliation: another Kingston settlement and a rural community 15 miles away. Meetings are held on neutral territory, at the Construction Resource and Development Centre offices. Adjusting to the changing market for building skills and the current decline in the construction industry, WCC has taken up new tasks such as house maintenance and repairs, along with widening its scope to extend services to other communities. Today, WCC is a registered non-profit company managing its own funds and affairs.

Source: Bertha Turner, Associated Housing Advisory Services, London.

BALDIA SOAKPIT PILOT PROJECT, KARACHI, PAKISTAN
Baldia Soakpit Pilot Project, a community-based sanitation initiative, can count its successes on two fronts. With sanitation as the starting point, health and the environment were improved for all local people. Equally important, a major breakthrough has been achieved by and for women in improving their own status and education, now and for the future.

Baldia: A typical Katchi Abadi
Two hundred thousand people live in Baldia Town's 28,000 households. From 1947 onwards, they came from rural villages, bringing with them the ties of

family, tribe, village, culture and their own traditional community organizations. They built their houses on illegally occupied land, with materials obtained locally, as do two-thirds of Karachi's population. About 30 similar self-built settlements, or Katchi Abadis, cluster around Karachi. Within 25 years, 87 per cent of Baldia's houses were improved from mud to concrete block walls. Water, once available for only one hour every two days, now flows for two hours a day, via standpipes.

Baldia's people work in low-paid, unskilled jobs. The average household of nine with two wage-earners earns just 700 rupees to provide the barest essentials. A quarter of the families earn less than the subsistence wage of 500 rupees.

Before the project started, the human excreta from bucket latrines inside the houses gushed through holes in the outside wall to be removed irregularly by poorly paid "sweepers". Urine and waste water simply ran down the unpaved street. Stench, child deaths and chronic diseases were rampant.

Improved soakpit design

The Baldia Project put forward the idea of low-cost, long-life soakpits in 1979. People's cultural and rural prejudices made them dubious about having both the system and their bodily wastes inside their homes. Soakpits are not new, but Baldia's had an improved design. A bucket of water could flush the pan – an important feature in an area with limited standpipe water. The 14-foot-deep pit can be used by the average family for 25 years before it needs emptying. Lining it with local stone or concrete blocks prevents collapse and liquid seeps out through tiny gaps left between the blocks. The pit tapers towards the top, reducing the size of the covering slab needed. Inside the house, the top closes with a latrine pan set in a concrete slab. Keeping the U-shaped bend of the pan filled with water seals off the underground area, reducing odours and flies.

Spreading the word on the doorstep

Technology cannot solve problems unless it is acceptable to local people. The project started with no office or vehicles, but with one woman community organizer from Karachi University's Department of Social Work, and one part-time engineer. For two years they focused on one area, walking from house to house, explaining the soakpit to people on their own doorsteps. They began to be trusted by local leaders and organizations. Finally, a few families decided to build their own soakpits. These demonstrations showed what could be done. After that, word of mouth took over and people enthusiastically took up the idea. Twenty other communities outside Baldia – over 40,000 people – followed in the next three years. Over 2,000 pit latrines were built by 12 existing community organizations. Gradually, a series of small projects has become a large-scale development.

Producing more for less

The project was a collaborative venture involving UNICEF, the Pakistan government and local communities. By 1984, 430 demonstration soakpits had been built by UNICEF trainees at the cost of 1,300 rupees each. An additional 200 pit latrines and 3,060 community-built soakpits followed. Baldia people managed to cut costs by more than half in the next round of 2,630 soakpits. They pared costs down further to 800 rupees each, when they built 4,000 more. By 1985, 26 of Baldia Town's neighbourhoods were doing their own sanitation. Karachi Metropolitan Council was even persuaded to surface streets and pavements, and to provide light, power and a better water supply by 1984.

Home schools

The difficulties of disseminating printed health information inspired another set of projects which may be even more significant. Illiteracy for Pakistani women is 78 per cent and for men 59 per cent. Girls are kept at home until marriage, and their education is discouraged. Women workers on the Baldia Project spent two long years meeting with men's groups before finally being allowed to speak with Baldia's women. Stimulated by the project, Baldia's young women were trained as teachers to run 107 schools in their homes. By 1985, 3,000 pupils were enrolled, 80 per cent of them young women whose families would not allow them to attend other schools. Teachers are paid by the community and have their own government-registered professional organization. A self-managed skills training centre also instructs 120 girls at a time, who then teach their skills to others, multiplying the numbers of skilled women.

Primary health care

Thirteen teachers also trained as health workers to service twelve new health centres. About 1,500 children were immunized, and over 1,000 mothers registered, with one-third trained in oral rehydration to prevent child deaths by dehydration. Growth monitoring raised 400 children above malnutrition level and disabled children received treatment. Three family-planning centres and a maternity home were built.

The Baldia Project shows that a woman community organizer can motivate both men and women, and pave the way for introducing new activities for women. Previously confined to domestic tasks, they now operate schools in their homes and conduct adult literacy classes outside Baldia. Women are learning and teaching each other new skills. A fundamental change is taking place between men and women and rigid customs have been relaxed. Men have begun to accept and to benefit from women's newly discovered capabilities, and the women feel more confident. Through this evolving relationship, their children will also benefit.

Source: Bertha Turner, Associated Housing Advisory Services, London.

URBAN AGRICULTURE IN LUSAKA, ZAMBIA

In the straitened economic circumstances of the 1980s, and especially in countries which are trying to cut down food imports and encourage agricultural production by increasing producer prices and eliminating urban food subsidies, it is sometimes argued that urban food production should be encouraged. Food has always been produced in Zambian cities, although growing maize in urban areas was forbidden and maize crops (seen as a health hazard because of their potential as a breeding ground for mosquitoes) were periodically destroyed. In Lusaka, the need to produce food has been used to justify the large size of housing plots, while the use of vacant land for rainy-season gardens has been tolerated.

Home plots and gardens

Low-income families cultivate plots next to their houses, or rainy-season gardens on vacant land at the fringe of residential areas; some cultivate both. In the squatter areas, relatively high densities prevent many residents from cultivating near their homes, but where plots are larger, for example in serviced settlement schemes or in the overspill areas, the proportion of households cultivating areas adjacent to their houses increases. A recent survey showed that 85 per cent of families grew food crops; 59 per cent cultivated a plot near their homes and 44 per cent had rainy-season gardens. Overall, more than half of Lusaka's low-income families grow food, and most of these are home owners rather than tenants. The size of gardens varies: cultivated plots near homes are smaller than rainy-season gardens.

Only a minority of families (about 15 per cent) with rainy-season gardens "buy" land by making small payments to previous users. The rest take over land which is given up by another family or cultivate the nearest available patch of unused land. Although cost of land, therefore, does not limit cultivation, and households spend relatively little on seeds or fertilizer, the physical availability of land is a constraint. Moreover, the distance which cultivators have to walk to their plots is often considerable and may be increasing as urban development continues.

The role of women

Most families who grow food do so because their incomes are low and vegetables are expensive. Certainly women with children, especially those with large families, find it necessary to grow food, although for many, it represents significant additional work. Most of the cultivation is done by women: in one survey of rainy-season gardens, wives were solely responsible for preparation of the ground in 44 per cent of cases, planting in 59 per cent, harvesting in 65 per cent and drying and storage of the produce in 82 per cent. Some husbands prepare the ground and sow and guard the crops, and some, along with children, assist with harvesting. Families spend over 18 hours a week cultivating a rainy season garden, but to that should be added walking time, making a total of more than 21 hours – which, over a growing season of six months, makes some 500 hours. Women spend some 7 hours per week on the plots near their homes.

Crops grown
Maize, beans, groundnuts and pumpkins are usually grown in the rainy-season gardens, for they need space, relatively little care, are relatively invulnerable to theft or cutting down by public health officials and can be harvested quickly. Plots near homes are commonly used for tomatoes, onions, rape and cabbages, all of which need regular watering, are more vulnerable to theft and can be harvested continuously; and for fruit trees, some of which, especially bananas, are grown on full-up pit latrines, and which have the added benefit of giving shade. These crops provide a valuable and sometimes crucial supplement to the family food supply, at least from February to May, and represent, over that time, a saving of perhaps 10 per cent of family income. Relatively few cultivators (no more than a fifth) are able to grow a surplus for sale, although in a 1980 survey two-thirds of respondents said they would like to. Proceeds from sales vary from one or two kwacha up to K50 or so in a season.

Human settlements
As the cost of food continues to increase, but urban incomes do not, technical assistance and educational programmes could help to raise yields by encouraging the use of appropriate and low-cost inputs, and by increasing awareness of the nutritional value of fruit and vegetables. In one of the areas included in the 1980 survey, a nutrition and gardening programme had been under way for a number of years, run by a local NGO, Human Settlements of Zambia. Here, a greater proportion of households cultivated gardens and nearly a third of the cultivators were able to sell their surplus produce. There was greater use of compost to increase yields, especially on the home plots. Families who felt optimistic about being able to keep their rainy-season gardens for the next five years spent more on inputs and were more likely to be long-term urban residents.

Increasing production
To buy inputs or build up soil quality by using compost, families must be sure of a reasonable degree of security of tenure. At present such security is available only on a household's own plot, or by negotiating with the City Council for access to unused peripheral land for rainy-season gardens. If access to vacant land becomes more difficult in future, low-income families, especially those with children or new to Lusaka, will be prevented from cultivating as freely as they have in the past.

One answer is for the government to provide the option of larger plots in its new serviced settlement schemes. But large plot sizes increase the costs of infrastructure, especially water supply and road access, and, if this increased cost is to be financed by residents, very poor families will effectively be excluded from the settlement schemes. Another alternative is to reduce plot sizes and provide land for gardens adjacent to low-income residential areas, as was tried in one earlier serviced plot scheme, and to provide them with water to allow for more intensive cultivation. Any formalization of urban gardening must, in addition, avoid erecting administrative hurdles to land access: individual control over cultivation decisions is crucial in enabling households to use food production successfully as part of their survival strategies.

Source: Carole Rakodi, University of Wales, Institute of Science and Technology.

WOMEN AND ENVIRONMENTAL CONSERVATION

PART II

WOMEN AND
ENVIRONMENTAL
CONSERVATION

CHAPTER SEVEN

Women Working for Conservation

The degree to which we all are involved in the control of the earth's life is just beginning to dawn on most of us, and it means another revolution in human thought (Lewis Thomas, *The Lives of a Cell*).

Changing people's awareness about the environment will indeed require a virtual revolution. Even more difficult, however, is changing attitudes towards women and the environment. We have seen the key role that women play as conservationists and sustainers of the environment. As long as their role remains informal, it is easy for governments, international agencies, and even NGOs to gloss over their importance, and to ignore their potential usefulness as a force for sustainable development.

This chapter documents the growing and increasingly organized nature of women's involvement in conservation. It shows that, where women receive support, they create active, dynamic and effective organizations. But, as the second half of this book argues, they do need support – in the form of training, education, family planning, and international assistance. By not acknowledging women's *de facto* role as conservationists and giving them support where possible, governments and international agencies risk losing a valuable ally in the fight to conserve the environment.

WHAT IS CONSERVATION?

In 1980 the *World Conservation Strategy* was published by the International Union for Conservation of Nature and Natural Resources (IUCN). The WCS, intended as an action plan for governments to develop their own national conservation policies within its guidelines, gives us probably our most useful definition of conservation. According to the WCS, conservation is:

"The management of human use of the biosphere so that it may yield the greatest sustainable benefit to present generations, while maintaining its potential to meet the needs and aspirations of future generations.... Thus conservation is positive, embracing preservation, maintenance, sustainable utilization, restoration, and enhancement of the natural environment." Conservation, like sustainable development, is not only relevant to nature itself, but also to people; "while development aims to achieve human goals largely through use of the biosphere, conservation aims to achieve them by ensuring that such use can continue" (IUCN, 1980).

"Conservation is a process, to be applied cross-sectorally, not an activity sector in its own right. In the case of sectors directly responsible for the management of living resources, conservation is that aspect of management which ensures that utilization is sustainable and which safeguards the ecological processes and diversity essential for the maintenance of the resources concerned. In the case of other sectors – such as health, industry and energy – conservation is that aspect of management which ensures that the fullest sustainable advantage is derived from the living resource basis and that activities are so located and conducted that the resource base is maintained" (IUCN, 1980).

Living-resource conservation, continues the WCS, has three specific objectives:

- to maintain essential ecological processes and life-support systems on which human survival and development depend;
- to preserve biological diversity (the range of biological and genetic material); and
- to ensure the sustainable utilization of species and ecosystems (IUCN, 1980).

Conservation and sustainable development are thus mutually dependent. The reliance of rural communities on living resources is direct and immediate. And unless those resources are conserved, there is no prospect of improving living standards. It would be wrong, however, to conclude that conservation is a sufficient response to the problems of the rural and urban poor. People whose very survival is precarious and whose prospects of even temporary prosperity are bleak cannot be expected to respond sympathetically to calls to subordinate their acute short-term needs to the possibility of longer-term returns. Conservation must therefore be combined with measures to meet short-term economic needs. Yet where they have the chance, women already adopt such a policy.

Woman walking to town – 15 years ago this area of the Sahel was thick with trees (*Jeremy Hartley/Oxfam*)

Planting millet in Burkina Faso (*Jeremy Hartley/Oxfam*)

Women and their daughters spend many hours every day collecting water
(*Jeremy Hartley/Oxfam*)

Carrying firewood to market in Nepal (*Caroline Penn/Format*)

Cooking is a major use of fuelwood (*G. Salkeld/Oxfam*)

(*top*) Urban living in Coelhas, Brazil (*C. Pearson/Oxfam*)
(*bottom*) Watering saplings in Rajasthan (*Sue Greig/Oxfam*)

Women's co-operative harvesting potatoes in Rwanda (*Vincent Banabakintu/Oxfam*)

Women's positive action

Taking each of the WCS objectives in turn, it is possible to assess the contribution that women are already making to sustainable development.

To maintain essential ecological processes and life-support systems

The maintenance of soil fertility and structure by soil regeneration and protection by forest conservation, water management and the prevention of water and air pollution are required to accomplish this goal.

As we have seen, women make up a large number of the world's agriculturists. In their day-to-day work, they draw on their traditional – and often extensive – knowledge of soil conservation to effect techniques such as terracing, crop-rotation and agroforestry. Many of these methods have been proven over hundreds of years. Locally women are involved in several soil-conservation projects, like those in Cape Verde and Nepal, but under most official programmes, women are not involved at all.

In local forestry and forest management, women are playing an increasingly important role. Many of these efforts, like those in Kenya by the Green Belt Movement and other organizations, have been described already in the chapter on forestry. Other examples are:

- The Fédération des Associations Féminines du Sénégal (FAFS) is supporting local initiatives of women to establish village forests either individually or collectively. It participates actively in national afforestation campaigns and organizes seminars and symposia on deforestation and afforestation.
- Following the devastating drought of 1982-4, the Association of Women's Clubs (AWC) in Zimbabwe and its members throughout the country came to realize the importance of maintaining their environment. AWC has embarked upon tree-planting projects in various areas in Zimbabwe, where drought-resistant indigenous trees are planted. The Forestry Commission supplies them with seedlings, while AWC solicits funds for tools and fences.
- In India, local forests established by women have a much higher survival rate after ten years than those which have been planted by the government: 80 per cent compared with 20 to 30 per cent.

In water management, too:

- One of the main goals of the Chipko activists and similar

organizations in the Himalayan region is to maintain the local water balance.

- The Sarvodaya Shramadana Sangamaya, a Sri Lankan development organization with a women's department, has initiated several efforts to improve the environment – in areas such as the protection and care of wells, the construction of toilets to improve sanitation, use of home refuse to make compost, cultivation of home gardens, and a fuel-conservation programme in which mud stoves have been introduced to conserve fuel.

And women are active in the prevention of pollution:

- In Malaysia, the Consumers' Association of Penang, which has a special women's wing, is fighting industrial pollution of rivers. There has been an alarming decrease in fish production, the serious effects of which have been felt in local communities, particularly by women.
- Women along the coastal areas of Papua New Guinea are starting to look after environments beside the sea where people collect sea salt for cooking.
- Women in Bhopal have joined forces to protest against Union Carbide operations since the disaster of 1984.
- In Quito, Ecuador, women have recently organized themselves to resist the emission of dangerous chemicals from industrial plants.

To preserve biological diversity
Women have a profound knowledge of the plants and animals in their living environment. Traditionally, they use a variety of indigenous species – trees and other plants, and even animals – and this multi-functional approach promotes the preservation of these species. In the case of seed selection for agricultural uses, for example, women seem to know exactly which seeds are best adapted to environmental conditions. One NGO, ENDA-Zimbabwe (Environment and Development in the Third World), has designed a traditional seeds project to support rural women in caring for the indigenous crop plants they use to ensure food self-sufficiency (see Chapter 2).

It is often women who recognize the importance of the preservation of indigenous tree species. This has been the case in places such as India, Senegal and Kenya. In this respect, professional researchers and planners stand to learn from local people, particularly from women.

To ensure the sustainable utilization of species and ecosystems
Earlier chapters have shown that the sustainability of development, including the careful use of ecosystems and species, is of great

importance to women, for the environment, with its natural resources, provides the basis for their lives.

In the following interview, María José Guazzelli gives one example of the way in which women are seeking the wise use of species and ecosystems – in this case, through the Ação Democrática Feminina Gaúcha (ADFG) in Brazil. The final statements of the workshops on Women and Environment, organized by the Environment Liaison Centre (ELC) in Nairobi, 1985, underscored the need for these efforts towards sustainability.

Women are not represented
The role that local women can play in conservation has been little recognized by environmental organizations. This is reflected in the very limited extent to which local women are involved in conservation projects, including those, for example, funded by the WWF and IUCN. Nor are they well represented in NGOs, although the record is better there.

In preparation for this book, we conducted an inquiry of 46 organizations in developing countries. Women made up less than half of the professional staff in 31 of them. In only nine did women make up more than half the staff members. Forty-two of the organizations sampled are headed by men and only four by women. In other international organizations their position in the structure and policies, as we discuss in Chapter 11, shows the same pattern of marginalization.

Although some 18 of the environmental NGOs responding to the survey have no working relationship with women's organizations at all, it is nevertheless encouraging that a majority (27) do collaborate with women's groups. This trend should grow as an increasing number of well-qualified and articulate women, like María José Guazzelli, Vandana Shiva and Shimwaayi Muntemba, speak out on the crucial importance of women's role in sustainable development.

INTERVIEWS

VANDANA SHIVA, BY RENU WADEHRA
Dr Vandana Shiva is the coordinator of the Indian Research Foundation for Science, Technology and Natural Resource Policy and of the South Asian Seeds Action Network and the World Rainforest Movement. She works closely with the

Chipko movement and other environment groups in rural India. As she says, her passion is to "de-professionalize herself, to learn from the people, especially from rural women, and to channel their knowledge into policy."

Q: Why did you get involved in women's issues?

A: Partly because I am a woman, and partly because of the invisible training my sister Mira and I received at home. My mother, in her time, was doing exactly what we are doing today: she gave us a certain confidence that is denied to women in contemporary society.

Q: You are actively involved with women on environmental issues; what evidence do you have that women themselves have taken action?

A: In one or two cases, as with the Chipko movement, their actions have shot into the media. But I know for certain, no matter where you go, that if there is a scarcity of water, women have protested; if there has been over-felling of trees, women have resisted it. I know that when milk is being commercialized, women have said that it must first be saved for their children, before it is sold. When commercial crops have been introduced, women have tried to protest. But these are invisible protests. They have stayed invisible and died out. Women's leadership has never been projected adequately, either in terms of the ideas they have stood for or in terms of the foundations for action they have laid. Women have conceptualized the ecological crisis. They have told me what bad development is about, and they must resist it.

Women in India have led on these issues for decades. Mira Behin articulated ecology way back in the 1940s. Sarla Behin was an early ecological activist: she nurtured many organizations which gave rise to the Chipko movement and inspired some of our later environmental activists like Shati Suresha Devi. To me, these are important connections: it turns out that women anticipated the men both in defining environmental problems and in fighting against them.

Q: How has the government conceived the idea of ecology and in particular the afforestation issue?

A: In my mind the only real conservationists can be those who live with the resource, who keep the resource alive and reproducing. That goes for forests, rivers, streams, wells. Unhappily, the government has not understood that it is at the local level that empowerment is needed and that it is essential to remove crippling forces such as the Forest and Land Acts, which force people to destroy natural resources. The government's forest policy is now being called a deforestation policy. There is much talk about the environment, but no talk about afforestation. And the centralized policy structures in Delhi do not allow the ordinary villager to function as a conservationist. There may be many children planting trees, but the villager, who is the only person in the long run who can make sure that India is green again, is being marginalized.

Q: We know that, in the past, women had more control over the forests. Now, many NGOs are involved. Do you think women are getting back their decision-making power?

A: The real issue is one of rights. Unless people have a right to resources, they cannot conserve them, and those rights have been denied since colonial times. We have male co-operatives to provide commercial timber to outside markets; now we have female co-operatives as well. But this is not recovering rights. This will only be the case when an ordinary village woman, the last rung on the ladder, is able to provide the basic needs for her children from the forest: then the rights she had earlier are restored. Until then, women have no control over the forest. For me, women's control is not in planting trees, but in making sure that any wealth generated is used by them for the survival of their children and their community.

Q: What do you understand by a woman's view of sustainable agriculture?

A: India is an ancient civilization which has had 40 centuries of experience in managing its resources, and the fact that it survived well shows that they were managed well. But imported development projects from Europe and North America have had a drastic impact on forests, agriculture and water resources. All three can be conserved under systems of sustainable agriculture, but the Green Revolution has brought discontinuity. When you say increase of wheat yields, you do not mention that in order to cultivate hybrid wheat, you are displacing a number of associated crops, including the wheat which was sound for the women. It was food for the family and meant employment. To me, sustainable agriculture is about identifying the genetic diversity, identifying what were the valuable species for nutrition and high yields. You will find that the sustainable agriculture package is a must and that it is based on the participation of women.

Q: All over the world women are facing environmental problems. In your view what would be the best way to create an international network so that they could learn from each other?

A: If somehow ecological stability has to be created, it is essential that those who are the victims of destruction have a means of expression in their hands. A network to me is a political system which is different from a political party. It is also different from NGOs because they can also centralize power and information into their hands and then, either willingly or unwillingly, distort it to suit occasions rather than to meet the needs of the people whom they represent. Once you have global destruction, the only countervailing force can be a global movement; somehow we need a process of networking that will allow all the local movements to link up with each other, but maintain the messages and the power of the grassroots. Maintaining that power and increasing it is the challenge. But we have problems of money and language. Somehow networking has to take place, but it is a very difficult task. I know how hard it is to get real people in touch with each other; there are too many intermediaries standing in the way of direct communication between grassroots organizations.

MARÍA JOSÉ GUAZZELLI, BRAZIL, BY IRENE DANKELMAN

María José Guazzelli is an agriculturist specializing in organic farming. She works as an agricultural advisor to private farms in the state of Rio-Grande do Sul in South Brazil, but her main work is with the Ação Democrática Feminina Gaúcha (ADFG), Amicos da Terra (Friends of the Earth) in Brazil, a non-governmental environment organization, for which she coordinates the Vacaria Project, including a demonstration farm and training programme on low external input agriculture (see page 24). The project involves working with peasants, extension workers, agronomy and veterinary students. She has written booklets on alternative technology in pest control and gardening.

My experience is that it is usually women who have been the first to organize and lead ecological grassroots movements to press for change, perhaps because we are directly involved with family affairs. The model of development proposed and executed by men has proved to be unsustainable, even if it can be economically successful in the short term. It is now, and will increasingly be, women who fight for sustainable development.

In my own work, I can see clearly how environmental degradation and unsustainable development have severely affected women.

Until the mid-1960s, the rural structure in Brazil permitted a reasonable standard of living for a peasant family: a piece of land, a house, enough food, and – most of the time – surplus food production, which could be sold. But from 1964 to 1985, the development system chosen by the Brazilian military government, with its increased industrialization of the country, has radically changed this situation. The Green Revolution brought increasing debt, through bank credits given to buy fertilizers and pesticides, heavy machinery and hybrid seeds. The technical and financial pressures to cultivate soybeans for export has ended in destruction of the environment, with land contaminated and eroded. And by concentrating land ownership in the hands of a few, the Green Revolution has promoted a rural exodus in which 30 million people are involved. Poor health, premature delinquency, and the birth of deformed children (because of the use of pesticides) affect women directly. Many turn to prostitution as the only way to survive.

The work we do in ADFG-FOE-Brazil can be seen as one example of how a women's group (with male members since 1983) can be extremely active in the field of environmental conservation. Magda Renner and Giselda Castro respectively the president and vice-president of ADFG, have been working in this area since 1974: fighting for social and political changes that promote sustainable development and, consequently, environmental conservation.

A message I would like to share with you is my idea that we – millions of people, especially Third World women – all need development and better living conditions. But development will only be socially right and sustainable in the long run if it includes respect for the environment and wise natural resource management as an essential part of decision-making.

SHIMWAAYI MUNTEMBA, ZAMBIA, BY IRENE DANKELMAN

Dr Shimwaayi Muntemba is director of the Environment Liaison Centre, Nairobi. Until November 1986, she was coordinator of a programme on Food Security, Agriculture, Forestry and Environment at the World Commission on Environment and Development (Brundtland Commission) in Geneva. The main focus of the programme was to relate the exploitation of agricultural resources, including forestry, to sustainable development. Among many issues addressed by the programme was the role of women in food production. She is the author of several books on agriculture and development.

Since 1980, I've been researching women and agricultural change. One aim has been to restate and demonstrate that women everywhere in Africa are in the vanguard of those trying to ensure food security because they are the main food producers. The critical questions are: What happens to the resource base and to food availability once the main production force, women, is dispossessed and undermined, as has been happening throughout the twentieth century? What happens when production is large-scale and no longer intimately linked to the resource base? How much hope remains for enduring food security if the main producers continue to be marginalized to the extent that they must depend on ever more fragile ecosystems? The task, then, is to suggest how to focus on women as a solution to rural underdevelopment, low agricultural productivity and food insecurity.

I am also concerned with the overall underdevelopment of rural areas. Here, the focus is on processes which undermine peasant agriculture and fisheries: the takeover of resources by large-scale producers; the failure to increase productivity, because of a lack of equipment, capital, markets and communication. These processes, particularly the first, have brought poverty for the majority of small-scale producers who, in order to survive, have no option but to exploit the fragile resource base in ways that undermine it further.

All this work has convinced me not only that environment is most linked with sustainable development, but also that women play a pivotal role in this, particularly in the area of food security. Everywhere, in both the developing and the industrialized world, women are concerned with the quality of food and water because, despite the strides women's liberation has made, they continue to shoulder the greater part of the family health burden. Women are thus directly concerned with human welfare, and this depends on sustainable development.

Women's relationship to the environment and their importance in sustainable development cannot be overstated. They are themselves painfully aware of the threat to their own lives and those of their children from onslaughts on the environment. In the developing world, a few organizations are responding specifically to the environmental threats, but the majority of women's organizations, although they do not seem to be dealing directly with the environment, do invariably concern themselves with the consequences of environmental changes: the Consumers' Association of Penang; the Working Women's Forum in India in their dealings with fisherwomen and milkmaids; the many Mandals (women's groups) in the Indian Himalayas; KENGO in Kenya; Fundación Natura in South America.

In Africa, agriculture-related cooperatives, clubs, associations and many other groups of village women all concern themselves primarily with the sustainability of resources. They struggle for the freedom to select a less fragile and more durable foundation on which to increase and sustain productivity. A successful outcome would ensure sustainable livelihoods and, with time, sustainable development.

To women's environmental organizations in the North my message is to:

- Bring pressure on donor agencies to ensure that in environmental matters, their targets will be mainly women.
- Be aware that environment is one area in which rural women are highly knowledgeable. Your approach whenever you work with women's organizations in the developing world should be one of dealing with those who know, but who lack power and financial resources. Environment can be more effectively managed if those who know it intimately are involved in both decision-making and the execution of projects.
- Encourage women and their organizations in the developing world to be confident of the power their knowledge confers on them and to use it to the full.

CHAPTER EIGHT

Training Women

Educate a boy and you educate one person.
Educate a girl and you educate a nation
(A.Ibn-Badis, Algerian Muslim reformist, 1889-1940).

Although education and training opportunities for women neither automatically improve their position in society nor guarantee their equal participation in development, inequality in education is a vital constraint on women's progress in both rural and urban areas (Ashworth, ed., 1981). Generally, women who are educated and trained improve their status within the family, the community and the nation. Evidence shows that educating girls may be one of the best investments a country can make.

WOMEN'S EDUCATION

Women's access to education has undoubtedly increased during the Women's Decade, yet they still make up two-thirds of the world's illiterates. In many countries, women complete the equivalent of only two or three years of primary schooling – and this with erratic attendance (World Bank, 1985). Enrolment ratios in primary and secondary schools show that girls have much less access to formal education than boys, who, in some countries, attend school ten times more frequently than girls of a similar age. The drop-out and wastage rates are usually over three times greater for girls than for boys. Girls are the first to be asked or forced to leave school if there is a financial crisis in the family.

For a variety of reasons, girls normally achieve lower grades than boys. In most cultures, they are not spared their household chores because of school attendance, so they are physically tired and often in a poor state to learn (Verghese, 1983). The environmental crisis further minimizes their educational chances, for, as earlier chapters have

Figure 6: *Women's education*

	Illiterates aged 15+		Enrolment ratio primary/secondary schools	
	% women	% men	% women	% men
Bangladesh	82	60	26	48
Egypt	78	46*	51	77
Guatemala	62	46	51	48

Source: UN Decade for Women, *Selected Indicators UN 1985*; figures for 1980–84 (*1975–79).

shown, environmental degradation – especially the loss of forests and deterioration of water resources – places a greater burden on girls than it does on boys.

The quality of education as well as access to it is important. It is by no means true that any kind of education is better than none. Formal teaching in schools and other institutions far too often reinforces the prevailing social values that label women as "second-class". It has been shown that even when girls form the majority of students in a mixed class, they rarely get more than a third of the teacher's attention. The reasons for discrimination against females are varied, ranging from parental suspicion of an unknown school to a vehement denial of the necessity of educating women. A further reason is the fear that education may bring about less compliant attitudes in girls (Verghese, 1983).

Women have the knowledge

Even though it may result in social disadvantages, a lack of formal education does not necessarily imply a lack of knowledge (Allison *et al.*, 1985).

In all societies, women bear the major responsibility for child-rearing. Whatever the family structure may be, they spend much more time at home than do men. Yet their household skills and their crucial role in child care are not recognized – and do not qualify them for a share of power and decision-making. Women's knowledge and skills as educators are also consistently undervalued. Yet women have always played the central role in informal education as custodians and transmitters of indigenous knowledge and culture. In all parts of the world, women have been involved in traditional medicine, farming and the processing and preservation of food. This knowledge is vital for the survival of families, communities and indeed the environment; it needs to be valued, protected and developed in parallel with the acquisition of new knowledge and skills from which women are still largely excluded.

Women are the first environmental educators. It is through contact with their mothers that young children first learn to see and understand what is happening around them, and begin to feel how they themselves are related to it. In the home, the groundwork is in place to demonstrate "connections". Children brought up learning "to connect" with their environment are more likely to "acquire in a formative age of imagination, a picture of themselves as citizens of a shared and indivisible planetary home" (Ward, 1973).

THE IMPORTANCE OF TRAINING

Training is a dynamic and adaptable tool of interaction and a powerful means of bringing women into the development process. Special training can revive and strengthen traditional skills and build upon their enormous fund of indigenous knowledge, so transforming it into the capacity for action. Training can also help to eliminate women's isolation and build confidence.

Most existing training, unhappily, falls far short of these possibilities. Often the only training to which women have access is in the so-called "feminine" occupations of health, nutrition, sewing, handicrafts, childcare and home economics. These skills, necessary though they are, do not enable women to participate equally in development. They have to be helped to develop a capacity for income earning, to be trained in leadership, decision-making and in new technologies. Their participation in these areas is still insignificant compared with the training offered to men. "Perhaps the most flagrant and best known dualism occurs in rural. . .agricultural training programmes in which women's meagre access to training in no way reflects the overwhelming percentage of time that they spend in agricultural labour" (Verghese, 1983).

Earlier chapters and case studies have shown how effective it can be to work with women's groups and other local NGOs to reach women and train more of them to participate in sound development projects. But the experience of INSTRAW (a case study in this chapter) is that attitudes towards the training of women will have to change, and that more resources and better coordination are needed.

Good practice

Most environmental training programmes – on soil regeneration, afforestation, energy saving, water management – still discriminate against women. But there are important exceptions to this practice.

The Centre for Human Development and Social Change in Madras, for example, annually conducts highly successful ecodevelopment camps for young people, especially women. Over ten days, the camp participants have lectures, discussions, film shows and tree-planting sessions. They visit other groups to see and learn about practical environmental action. In the villages of the Ivory Coast, the African Institute for Economic and Social Development organizes special training meetings for women's groups on such issues as afforestation, soil protection and the use of organic manures. And in Senegal, AFOTEC is helping women to learn from each other (see below).

Good training can benefit women as well as the environment. In the Indian tribal areas of Chotangapur, poverty forces rural women into headloading fuelwood. This further exacerbates deforestation. The Xavier Institute of Social Service tries to address this problem by generating training projects which create alternative sources of income for women. A field project in Angora, headed by a tribal woman who is a postgraduate in rural development, is already under way.

In Botswana, the Mennonite Ministries have women working in the remote villages of Etsha and Tsaure to help local women improve their literacy and make more progress in horticulture, the care of livestock, and the growing of indigenous food crops. Training such as this gives women greater independence from food relief programmes and development projects which promote high-input agriculture.

CASE STUDIES

TRAINING AND TECHNOLOGY IN SENEGAL

AFOTEC, the International Service for the Support of Training and Technologies in West Africa/Sahel, based in Dakar, Senegal, is improving women's access to appropriate technology, credit and health services. Its aim is to strengthen the abilities of community groups, especially those which include women as active members, to organize and complete their own development projects. In the process, AFOTEC identifies specific and appropriate means to lessen women's workloads using locally available materials and human resources wherever possible.

What sets AFOTEC apart from most development organizations is that it uses indigenous techniques and methods perfected in Africa. To this end, AFOTEC continually updates its knowledge of practical innovations in afforestation, water resource management, agriculture, public health, credit, literacy and income generation. Through its Community Advisors Committee, composed of women and men chosen from communities where the organization is active, AFOTEC explores ways of fostering the exchange of technical information and everyday experience among rural populations and between ordinary people, researchers and technicians.

AFOTEC itself does not implement projects. Instead, community groups set goals, devise strategies and carry out the work. AFOTEC helps them to identify the resources they need and obtain the tools and materials or the training they do not have.

A major part of AFOTEC's activity centres on improving access to clean water. When a women's group needed assistance to get water from sources near their villages, for example, AFOTEC contacted an appropriate technology institute in Bamako, Mali, that had developed a simple, sturdy pump. Eighteen other women's groups in the region had expressed similar concerns about water, and they elected leaders who travelled together to Bamako. There they learned to construct and maintain pumps and returned to their villages with complete units. Six months later, after the participants had attended a refresher course, they trained other community members in pump construction and maintenance. Local Ministry of Health workers are now involved in helping these communities to keep their wells and pumping systems clean and well maintained.

AFOTEC has advised numerous groups on the different types of containers in which water can be stored without becoming contaminated. For example, a ceramic container which women use to collect water can be constructed with a small bronze tap built into the base. Instead of dipping their cups into the large container, thereby contaminating its contents, family members are encouraged to use the tap.

On a smaller scale, AFOTEC is piloting a project in which traditional doctors are encouraged to purify the water they use to make medicines. A creative Ministry of Health employee collaborated with a local blacksmith to devise an easy-to-use water distillery. Under the auspices of AFOTEC, he now demonstrates its use in Senegalese villages.

In other ways, too, AFOTEC promotes integrated development. A women's group in Senegal, for example, expressed interest in storing vegetables to sell during the dry season when prices were high. AFOTEC was able to describe a system of seed and vegetable storage used in Burkina Faso. Elected group leaders travelled to Burkina Faso, where villagers demonstrated their preservation method, which was easily adaptable to Senegal. The new methods led to increased profits for the Senegal group.

To maintain the momentum of community commitment to development, AFOTEC has established a flexible fund called "FONDAP" (Fonds d'Appui aux Groupements). Recognizing that results are often slow, especially in land regeneration, reforestation and preventive health care, AFOTEC allows participating groups to have access to FONDAP for several years. The fund is used to pay for training and necessary materials and implements – a crucial form of support if local initiative is to continue.

AFOTEC's successes stem from its conviction that indigenous African techniques and implements provide the most effective answer to problems in both villages and cities. By bringing knowledge from one African country to another, or from one village to another, AFOTEC ensures that development projects are appropriate as well as affordable.

Source: Non-Governmental Liaison Service, UN (1987), *Case Studies from Africa*.

INSTRAW

The United Nations International Research and Training Institute for the Advancement of Women (INSTRAW) was created to ensure the integration of women as participants in the development process. INSTRAW's work reflects the growing demand for concrete action and practical measures. Making links between research, information and training is the basis for all INSTRAW's programmes. Spreading information through networking and documentation exchange is fundamental to its work. Most of the institute's resources are devoted to training and advisory services in its work with development agencies and women's groups. It strengthens existing programmes by developing cooperative arrangements with training institutions. Not only does INSTRAW promote conventional training methods such as workshops and seminars, but it also designs innovative training schemes, using many different media and is involved in the preparation of university curricula on the issues of women and development. In addition, INSTRAW undertakes research on, and policy analyses of, women and their contributions to development.

INSTRAW has worked with the International Labour Office in Turin to produce two multi-media training packs on "women, water supply and sanitation", focusing on examples from Africa, Asia, the Caribbean and Latin America. The approach is action-oriented: participants are encouraged to identify possible solutions to the constraints they face in bringing about more effective water supply and sanitation systems. The aim is to use the training packs to promote a much greater involvement of women in the planning and implementation of water projects. Participants in these training programmes get seminars, audio-visual presentations, lectures, group work and study visits, and the packs are supported with additional materials in English, French and

Spanish.

Another INSTRAW objective is to establish "women in development" as a subject of study in formal and informal education and training programmes around the world. In the first phase of this work, the institute is conducting a survey of academic and other centres that offer courses on women-related issues. Analysis of the content of these courses will take place in the second stage of the work; a final phase will involve the design of new courses, targeted especially at university teachers.

Source: Dunja Pastizzi-Ferencic, Director of INSTRAW.

PAKISTANI WOMEN VISIT INDIA'S ENVIRONMENTAL NGOs

From 4 to 18 March 1987, a group of seven Pakistani women visited India to meet with Indian environmentalists. They were a diverse group: environmentalists, journalists, a teacher, development planners and field workers, drawn together on an IUCN-supported study tour. The group looked for trends which might indicate why the Chipko movement had been so successful in India, and why such a movement had not arisen in Pakistan.

Of particular interest to the group was the question of women's roles in Chipko. The movement is outstanding for its grassroots, widespread women's involvement. But how great a part did Chipko women play in decision-making? To what degree did they fill leadership positions?

Over two weeks, the Pakistani women travelled through most of northern India. They visited NGOs, institutes, community organizations and sites of major Chipko campaigning, including Nahin Kala, where Chipko people resisted limestone strip mining in 1986, and Tehri, a deforested area where there has been persistent opposition to the Tehri Dam. One of the leaders of the Chipko movement, Sunderlal Bahugana, told them about the new eco-development camps. And at the Laxmi Ashram, they saw the visible benefits of education for local girls and women. As the study tour later reported:

> The ashram has expanded into playing a role in environmental activities, and groups of girls are now sent to educate the villagers on environmental matters, teaching them which trees to plant, how not to deforest their habitat and, through the medium of drama and song, how to raise environmental awareness (Marker, 1987).

"For they are our life," run the lyrics of one such song, explaining why villagers should protect the forests, birds, and mountain wildlife.

The group's findings were at the same time sobering and encouraging. It was concluded that movements such as Chipko cannot be transplanted to another culture: they are too much a product of local circumstances to be neatly replicated elsewhere. Yet the lessons of Chipko, the Pakistani women found, are applicable everywhere: the importance of drawing on the strengths of the community, the value of training, and the need to use indigenous species of plants and animals. A second conclusion was that education plays a key role in determining leadership. The leaders of the Chipko movement are mainly educated men, whereas the women – most of whom are not trained – form the strength of the resistance.

After their tour the participants began to integrate environmental concerns in some way or another into their work: in education and teaching, in health care, by writing articles, and by starting environmental projects. Project proposals were drawn up from the team members' personal reports and from the ideas resulting from meetings with the Indians. A common conclusion was that too little was known about the Pakistani environment. Among the project ideas to emerge, therefore, are plans for a study of natural resource management and the role of women in the northern area of Pakistan, followed by a workshop, an historical study and analysis of environmental trends, an inventory of indigenous sustainable methods of environmental management, the drafting of a Pakistan's State of the Environment Report, and an afforestation project in Gunyar.

The "hidden agenda" of the tour – "to build a nucleus of women professionals for further environmental activities in Pakistan" – has been successfully completed.

Source: Aban Marker Kabraji, IUCN Pakistan.

CHAPTER NINE

Planning The Family: A Woman's Choice?

In 1987, world population will pass five billion. It is growing at the rate of approximately a billion people every 12 years. Every minute it grows by 150; every day by 220,000; every year by over 80 million. Ninety percent of this growth is in developing countries.... Is reaching 5,000 million a triumph for humanity or a threat to its future? (Salas, 1987).

If we lived in an ideal world, in which everyone – North and South, rich and poor – had access to the same amount and quality of resources and consumed as much (or as little) as anyone else; in which sustainable management of the natural world was fully integrated with development; in which the carrying capacity of land was recognized and honoured; in which appropriate technologies were available to all and practised widely, then the size of the world's population would be of no great concern. Indeed, a 1983 FAO study concluded that with more irrigation and the wise application of other known technologies, the earth could produce enough food for everyone. And, if the rich consumed less, environmental degradation would be reduced.

But today's reality shows that society and the environment fall far short of this ideal. An enormous gap exists in consumption rates, in the resources available and in the ways in which those resources are managed between richer and poorer countries, and between the wealthy and the destitute within the same countries. In these circumstances, rapid population growth in the poorer countries exacerbates the problems of survival and accelerates the rate of resource degradation – and in consequence, the burdens women must face.

When the present developed world passed through its demographic transition in the nineteenth century, "surplus" populations could be exported to relatively unpopulated countries. Today, no such solution exists for the people of the South, whose countries have an average growth rate of 2.5 per cent per year. With a projected world population in excess of 6,000 million by the end of the century, and 90 per cent of the growth taking place in developing countries, the struggle for survival of those who depend upon natural resources of food, fuel, water and building materials seems doomed to fail.

Overall, reducing population pressures on the environment is not simply a matter of limiting the numbers of those living in developing countries: major changes in development priorities are needed, as this book argues in other chapters. Population control is one element of a complex web of radical decisions that are required to bring about sustainable development. But "present rates of population growth", argues the World Commission on Environment and Development, "cannot continue."

> They already compromise many governments' abilities to provide education, health care, food security ... and ... raise living standards. This gap between numbers and resources is all the more compelling because so much of the population growth is concentrated in low-income countries, ecologically disadvantaged regions and poor households.... Governments must work on several fronts – to limit population growth, to control the impact ... on resources, and with increasing knowledge, enlarge their range and improve their productivity; to realize human potential so that people can better husband and use resources; and to provide people with forms of social security other than large numbers of children.... Giving people the means to choose the size of their families is not just a method of keeping population in balance with resources; it is a way of assuring – especially for women – the basic human right of self-determination (WCED, 1987).

THE CONSEQUENCES FOR WOMEN

"Women realize better than anybody else what an accelerated population growth rate means," says Makwavarara (1986:23), and certainly many Third World women do appreciate the problem. But they are victims of a harsh reality which militates against population control. Many Third World children do not survive to be adults; women are often left as sole providers for the household, with only their family to care for them in old age. Cultural and social values encourage large families. More children mean an extra work force and insurance for the future. As long as child mortality is high, the incentive to have more remains strong.

Increasingly, women in the South are acknowledging the heavy toll that continued childbirth and child-rearing exacts on their own and their children's health. Induced abortions which follow unwanted pregnancies – most taking place illegally and in unsanitary conditions – expose women to severe physical risks, and can permanently damage their mental health. It is estimated that at least 30,000 women in 1987 will not survive the experience, almost all of them in the Third World. Some demographers put the figure several times higher (UNFPA sources).

Pregnancy-related maternal and infant deaths are still unacceptably high in developing countries and could be reduced by family planning (IPPF, 1985-1). One study, based on the World Fertility Survey, which collected data from 29 developing countries between 1974 and 1984, and from interviews with some 150,000 women of reproductive years, suggested that spacing all pregnancies at a minimum of two years would prevent 500,000 infant deaths each year in those countries (IPPF, 1987). Dr Fred Sai, Population Adviser to the World Bank, goes further. He estimates that family planning alone could save at least five million children and 200,000 maternal lives each year by helping couples to space their children and avoid high-risk pregnancies.

INTERNATIONAL ACTION

International assistance on population through the United Nations Fund for Population Activities (UNFPA) is a relatively recent develop-ment. Since its founding, the Fund has been mainly involved in demographic trends and projections. Three important principles guide its work:

- the principle of national sovereignty in population matters;
- the provision of necessary information and services to individuals and couples to determine freely and responsibly the number and spacing of their children; and
- the notion that population goals and policies are integral parts of socio-economic development (Salas, 1986-2).

The International Conference on Population in Mexico in 1984 emphasized that it remains the free decision of couples themselves how many children they raise. Women, however, often have little say in the matter. The Mexico Conference recommended that

Swift action must be taken to assist women in attaining full equality with men in the social, political and economic life of their countries. To achieve this goal, it is necessary for men and women to share, jointly, responsibilities in areas such as family life, child care and family planning (Mexico City Declaration on Population and Development, August 1984).

The success of local projects which promote family planning along with health and child care shows that women *do* want to space their families better, and so reduce the number of children. This enhances their capacity to play a proper role in their local communities and to contribute to sustainable development.

But access to family-planning information and supplies – even for those who want them – is woefully inadequate. According to World Health Organization estimates, 300 million couples who want no more children still have no access to family-planning services. There are, moreover, powerful pro-natalist forces which seek to undermine assistance programmes for family planning. And women can suffer in other ways: there is continued evidence of iniquitous practices in the distribution of unsafe and inappropriate contraceptives, with Third World women being used as "guinea pigs" for new product testing.

An integrated approach
Women's development, education and emancipation are key factors in increasing the acceptance of family planning. The need for, and the viability of, an integrated approach which links family planning to other aspects of development – including improvements in income, health, nutrition and education – has been well demonstrated. Women have to identify their own priorities and must be enabled to participate at all stages of a project – planning, implementation and assessment (Huston, 1978 and see case study). The provision of information about family-planning methods, and subsidized supplies, is just one part of this effort.

One of the most experienced organizations offering family-planning information and services is the International Planned Parenthood Federation (IPPF) with its member Family Planning Associations in more than 120 countries. In 1976, IPPF initiated the worldwide "Planned Parenthood and Women's Development" programme, focused on women in developing countries and with special reference to the poor in urban and rural areas who are often not reached by conventional development programmes. The objectives are:

- to enable women to work together
- to teach them skills through training
- to improve the status of women

- to improve the welfare of families through increasing family-planning knowledge and practice (IPPF, 1985–2).

Many development activities have been included in the IPPF programme, ranging from income generation to the provision of community services, health care, skills and leadership training, the management of women's groups and responsible parenthood. Importantly, as the following case study shows, environmental action has been successfully incorporated into this integrated approach.

An example from Nepal

More and more evidence is emerging of the usefulness of combining family planning with environmental action. The Family Planning Association of Nepal, for example, is coordinating a long-term project to help local people improve their self-sufficiency and reduce their birth rate. During 1987, some 150,000 fodder trees were planted by small farmers in the Sindhupalchowk District, north west of Kathmandu, and some 10,000 couples in the same area were protected against unwanted pregnancies. Sixty per cent of people there live below the poverty line and, with over 11 people per cultivated hectare, the district has one of the highest population densities in the world. Poor diets are combined with a lack of sanitation, bad housing and polluted water. Only 15 per cent of the people are literate.

The project began in 1973 with the provision of basic health care and family planning, but expanded (with support from the US NGO World Neighbors and from OXFAM, UK) to promote increased agricultural productivity and family income by demonstrating improved fodder trees and soil conservation measures. Forests were disappearing through the collection of fodder in the dry season, and soil destruction followed. Now, a fast-growing Leuceana species, the "Ipil" tree, has been planted on the face of terraces, reducing soil erosion and allowing farmers to coppice the trees for fodder all year round. Stimulated by the project, local people have made other farming improvements and begun to accept family planning. This area now has one of the highest contraceptive use rates in Nepal and the birth rate is almost half the national average (Hamand, 1987). Not only the "hardware" of family planning, but a certain level of development, are necessary if couples are to be able to have fewer children.

WAYS FORWARD

The first "Population and Environment" agreement, signed by IPPF and IUCN in 1983, could be a good starting point for further activities of this kind. IUCN, in cooperation with IPPF and other agencies, intends to develop a comprehensive policy on population and natural-resource management, promoting the ideas in national conservation strategies and through other IUCN projects (IUCN, 1984). In 1987, IUCN began a Population and Sustainable Development Programme and is updating the *World Conservation Strategy* to take account of population issues.

In October 1987, seven agencies concerned with the effects of population growth upon women co-sponsored the Conference on Better Health of Women and Children through Family Planning. The World Health Organization, UNICEF, World Bank, UNDP, UNFPA, the Population Council and IPPF all agreed on seven steps forward to improve the position of women in this area:

- family planning must be included in all primary health-care services;
- progress on child survival is essential for successful family planning;
- the special needs of young people in the field of reproductive health must be addressed effectively;
- problems of unwanted pregnancies and induced abortion must be discussed and resolved openly;
- even though financial resources are constrained, it is possible and essential to improve the quality of family-planning and primary health-care services;
- freedom and equality for women, along with substantial improvements in their roles and status, are both necessary conditions for achieving these changes; and
- inter-agency collaboration must continue so that these ideas can be implemented at the country level (IPPF sources).

CASE STUDIES

PLANNED PARENTHOOD AND WOMEN'S DEVELOPMENT
Working together with women's organizations, the International Planned

Parenthood Federation (IPPF) has sought to respond to the needs identified by local women through its Planned Parenthood and Women's Development programme. Although this is made up of a great variety of activities, its salient aspects are family planning and the environment.

In Indonesia, the IPPF's member family planning association has worked with women's groups to increase incomes and promote education, health, and community development. Invariably, family-planning information and services and local environmental action have formed part of the work of these groups.

In 1977, the Indonesian Planned Parenthood Association (IPPA) and the Indonesian Women's Association, Perkumpulan Wanita Indonesia (PERWARI), began a joint project to provide training, management, technical and funding support to village women's groups to enable them to undertake development activities based on community priorities. These did not, at first, include environmental action. But as the project progressed, a majority of village groups began a range of environmental improvements.

Following the establishment of a joint planning and management committee, IPPA and PERWARI sited the project in West Java; by the early 1980s its activities had expanded to districts in Yogyakarta, Medan, South Sulawesi and South Kalimantan. In each of these areas, the approach was to identify suitable project districts, interested village women's groups and leaders who could be trained. A key element of these village consultations was the discussion of local priorities. Initial plans were always modified by the women's groups, even after the project began, as they discussed and refined their needs.

The focus of the project became the training of women's leaders in project development and management and in leadership skills. Following initial district-level training, women's leaders went on to train village women and assisted in the development of plans for community activity. District leaders kept in regular touch with the local groups, helping them with further training, problem solving and the identification of technical needs. Throughout this work, the local IPPA and PERWARI branches provided management and technical support for income-generation activities, literacy and other educational work, health and family-planning services. The IPPA made small sums available to the village groups as a revolving loan fund to enable individuals and groups to begin new enterprises such as tailoring, weaving, and food production based upon rice growing, poultry rearing and fish ponds.

Although specific activities varied according to local priorities, a common pattern emerged. For most women's groups, income generation became a major focus for their efforts – reflecting their need both to contribute to family finances and to have control, however limited, over their own money. Few of the project participants had received any primary schooling: literacy and numeracy classes were therefore important for individuals and for groups as a whole, to enable them to improve their own management. Classes generally included nutrition, health, child care and family planning. And there were group recreational activities – dancing, music and readings from the Koran. After a year or two, the need to formalize income generation and financial arrangements often led to the registration of the group as a cooperative, the opening of a bank account and the establishment of individual and group savings and credit schemes.

From the beginning, the women chose activities which would benefit themselves, their families and the community. In the identification of participating villages, IPPA, PERWARI and the district women's leaders were always careful to consult local community leaders, both men and women. These leaders were generally supportive, playing a key role in gaining the acceptance of husbands and the wider community – support that was crucial both at the community level, where women were forging more independent and assertive roles for themselves, and in the family, where women were seeking a stronger voice in decisions, including those about family planning.

For many groups, environmental action subsequently became an important part of their programme. This involved community "clean-up" campaigns, the digging of latrines, tree planting and educational work to avoid the contamination of drinking water. The women's groups often became a major force for collective action in their communities, especially on environmental activities. For example, often the whole village would take part in a "clean-up" organized by the women. Many groups were especially effective in reaching out to young people, and this led to the setting up of a number of associated youth development groups which took on environmental tasks.

The pyramid structure of this project enabled large-scale expansion to happen quickly. By the early 1980s, hundreds of village women's groups had been trained and assisted in the project, and over 4,000 women had participated in local groups. Not all were equally active or successful, but most were reported to have worked together for at least three years.

A project assessment in 1985 identified a number of key factors explaining the success – and problems – of this joint IPPA/PERWARI venture. A useful starting point was the existing collaboration of IPPA and PERWARI and the fact that both organizations were community based and had branch structures, which allowed a common approach. The initiative was promoted as the women's own project, not something imposed from an outside agency. Even so, extensive support from the IPPA and PERWARI headquarters and branches was crucial in providing the necessary training, management, technical and financial help needed by the women's groups, especially in the early stages. The project enjoyed good inter-agency support at the district and village levels, and government cooperation.

The organization of income-generating activities remains a major concern; continuous effort will be needed to ensure that proper marketing and business management advice is available before an activity begins. Success is also dependent upon good local leadership: there were sometimes problems when a change of leadership, or the sharing of responsibilities, was resisted. Finally, the rapid expansion of the project brought its own tensions: both IPPA and PERWARI branches were periodically overstretched.

Since the beginning of this decade, IPPA and PERWARI have continued to assist the women's groups and the revolving loan fund set up for the original participants is now being used to assist others. Funding received from other donors has allowed the project to expand into other parts of Indonesia.

Common lessons

Within the IPPF's Planned Parenthood and Women's Development programme a number of other community projects have integrated environmental action with family planning and women's development. For more than 20 years the "Mothers' Clubs" of the Planned Parenthood Federation of Korea have been involved in similar activities. Now incorporated within the Korean rural development programme as the Samuel Women's Association, these groups have become a potent national force, mobilizing community energies for development.

Elsewhere, the Family Planning Organization of the Philippines has a programme of training and financial support for women's development groups which integrate family planning with environmental action. Although they are most developed in South-East Asia, similar projects are under way in India, Bangladesh, Pakistan, Kenya, Egypt and Nicaragua.

All these efforts illustrate the determination of village women to work together on their own development priorities. The integration of women's development, family planning and environmental concerns is seen as natural and obvious at the community level.

Source: Nancy Fee, Programme Adviser, Women's Development, International Planned Parenthood Federation, London.

CHAPTER TEN
Women Organize Themselves

We are women with different needs, interests, experiences and concerns.
We are women organizing within our unique social, economic, political
context. We are part of a worldwide movement of women. And within our
diversity lies our strength (IWTC, 1984: 5).

Women have organized themselves for centuries, around village wells,
in tribal councils, and in other traditional, informal ways. But in the
past fifteen years, women have increasingly organized themselves in
forums that explicitly articulate their concerns, both at the grassroots
level and within international networks. Like any other disadvantaged
group, women organize because it provides them with a larger pool of
resources upon which to draw. It breaks down isolation which
otherwise would divide one from the other. Organizing gives women
the collective strength to achieve their common objectives (IWTC,
1984).

This chapter reviews some of the ways in which women have
organized internationally, including Forum '85, the International
Women's Tribune Centre (IWTC) and ISIS. Specific reference is made
to women organizing for environmental conservation and management
through, for example, WorldWIDE and the Y's EYES Network. The
case studies give some examples of local women's groups.

FORUM '85

At the end of the International Decade for Women, a non-
governmental gathering known as "Forum '85", was held between 10
and 19 July 1985 in Nairobi, in conjunction with the UN World
Conference on Women. At the Women's Conferences in Mexico in
1975 and in Copenhagen in 1980, similar NGO events had been
organized, but this Forum was the climax of a ten-year process – a
summary of the decade's insights and activities.

The Nairobi Forum was not a conference in the usual sense of the word. It brought together about 15,000 women from more than 150 countries all over the world, and was open to all, regardless of their gender, race or creed. The programme incorporated as many different activities as the space permitted: 125 daily workshops, informal meetings, networking, demonstrations, and individual contacts covering the themes of development, equality, peace, health, employment, education, family, new techniques and networking. No formal resolutions or declarations were made in the name of the Forum, but out of the workshops, recommendations were put forward to the UN Conference, which began several days later.

The Nairobi-based Environment Liaison Centre organized a series of workshops on women, environment and sustainable development. INSTRAW and UNICEF held one on women and water management. There were also workshops on women and bio-energy, organized by the Women and Energy Committee of Nairobi, on sustainable agriculture, organized by the Women's Revolutionary Socialist Movement of Guyana, and on food and water, organized by the World Council of Churches. KENGO, the Kenya Energy Non-Governmental Organization, set up a workshop on the importance of indigenous plants. The Green Belt Movement (see case study) also held daily displays and demonstrations, organized visits to Green Belt projects and mounted a successful tree-planting drive among Forum participants.

> The common denominator for the majority of the participants was the sense that an extraordinary energy had been released in Nairobi. The aura of equalness, of power, of limitless possibility, of new understanding and fresh perceptions energized the Great Court of the University (where the Forum was held) and all the women who crossed it daily (Forum '85:23).

> In retrospect, Forum '85 seemed the result of a huge explosion of energy – an event greater, indeed much greater, than the sum of its parts, spontaneous combustion fuelled the very atmosphere of Nairobi, fed by the mix of energy of women from all parts of the world and all countries (Forum '85, 1985:10).

As Dr Eddat Gachukia expressed it: "Women have planted a seed in Nairobi that will germinate and grow with the years into a forest. The achievements of Forum '85 will become apparent and grow increasingly strong as the years go by." A decade had ended. But for many women, a new decade had begun in Nairobi. Each woman would return from the city carrying that collective energy back to her own community.

INTERNATIONAL WOMEN'S TRIBUNE CENTRE

The International Women's Tribune Centre (IWTC) was founded shortly after the 1975 non-governmental world meeting of the United Nations Decade for Women. Since 1976, IWTC has worked, through a three-part programme of technical assistance, reference and referral, to support a network of some 13,000 individuals and groups involved in women's projects and development assistance activities in the Third World. Its small staff, working with different groups and consultants from the developing world, produces highly visual, low-cost information and training materials to support the development of participation and leadership skills so that women may play a more active part in the life of their communities. The organization produces, for example, quarterly newsletters, regional resource manuals, training manuals, information bulletins, annotated bibliographies and contact lists.

IWTC has a strong commitment to the vital role women's organizations play as advocates, activists, providers of direct services to disadvantaged groups in their own countries, and as creators of alternative approaches to development. As such, the Tribune Centre responds to requests for collaboration and technical assistance from women's groups throughout the developing world.

ISIS/WICCE

Feminism means attitudes and behaviour which are consistent with the beliefs that men and women are equal in their rights and responsibilities and can exist and work together in harmony with each other and their environment (Louanne, ISIS, 1986).

Feminism is a philosophy going beyond limitations dividing all living beings on this planet (Monika, ISIS, 1986).

Feminism is a system which is: egalitarian (i.e. non-patriarchal, non-hierarchical and non-oppressive); democratic; allows for individual creativity and the full development of all individuals; has a consciousness of the importance of preserving the environment (Valsa, ISIS, 1986).

ISIS, the Women's International Information and Communication Service, was established in 1974 by a collective of women. Its main objectives are to gather materials from local women's groups and the feminist movement and to make these resources available to other women. With an expanding network of over 10,000 contacts in 150 countries, ISIS International in both Rome and Santiago, Chile, provides information and referral services, produces regular publications in both English and Spanish, coordinates regional and international networks, and offers training and technical assistance in communication and information skills.

ISIS also coordinates the International Feminist Network (IFN), a communication channel through which international support and solidarity can be mobilized for women's struggles. The IFN is used to spread information about international women's actions and demonstrations.

The *ISIS International Women's Journal* is produced by Third World women's groups and looks in detail at issues around which women are organizing. Its annual *World Newsletter* focuses on specific themes such as appropriate technology (1986) and new technology (1987). The *Resource Guides* provide an overview of organizations and literature dealing with issues such as women and development.

One of its most interesting projects is the Women's International Cross-Cultural Exchange (WICCE) programme, operated from the Geneva office. Women from around the world participate in an exchange for several months with women's organizations in other Third World countries.

DEVELOPMENT ALTERNATIVES WITH WOMEN FOR A NEW ERA (DAWN)

DAWN was launched as a Third World initiative in 1984 in Bangalore, India. Members of the group came from Asia, Africa, Latin America, the Pacific and the Caribbean countries. Several Third World institutions are involved in DAWN, which now has contacts the world over, including some in Western countries.

DAWN's main purpose is to mobilize opinion and to create a global support network for equitable development. An immediate aim is to establish guidelines for development which can be used by the network as a base for action. The organization poses serious questions about the content and style of present development. For women, argues DAWN, the "growth-centred" approach to development usually means reduced or, at best, slow-growing access to resources and jobs, trade-offs between employment and wages or working conditions, and increased work burdens in subsistence activities and in reproductive tasks. Both the monetary and economic crisis, as well as major environmental, demographic and technological problems, have serious effects on agricultural land, food and natural resources availability. Unemployment and under-employment, urbanization, migration and the increasing burdens of food production have been particularly severe for women, especially poor women.

DAWN's concept of development is, first and foremost, "people-centred". The values which sustain this vision are those that are widely held in the international women's movement, namely: cooperation, sharing, responsibility for others, accountability, resistance to hierarchies and commitment to peace. The issues which DAWN addresses are:

- the control and distribution of societal resources;
- sustainable management of natural resources and environments;
- the control of demographic pressures through women-centred approaches;
- the restructuring of power between the sexes, within families, communities and society.

WORLDWIDE

Women need a global village well around which to gather and share information about the environmental systems upon which many are profoundly dependent. Women must be willing to learn from each other so as to serve as voices for others who are voiceless. Women need to know that they are not alone in their attempts to salvage the earth for the future. Women need to join their healing hands (J. Martin-Brown, *WorldWIDE News, 1986, 1:4*).

WorldWIDE – World Women Working for Women Dedicated to the Environment – was founded in 1982 to:

- educate the public and policy-makers about the effects, specifically on women, of the destruction and contamination of natural resources and ecological systems, and to present women's perspectives on these issues;

- increase women's involvement in the design and implementation of development policies;
- support and expand the influence of women in organizations engaged in environmental and natural resource activities;
- encourage women, individually and collectively, to take up environmental and natural resource management goals.

WorldWIDE has an International Advisory Council with representatives from the major global regions. Several members are also Senior Women Advisers on Sustainable Development to UNEP, and on WorldWIDE's board of directors sit representatives from Latin America, the USA, India, Egypt, Ghana, Ecuador and Kenya.

To decentralize its activities, WorldWIDE has created local Forums analogous to chapters of a global organization, each having its own president and officers under the terms of WorldWIDE's by-laws. Each Forum arranges its own agenda and events and is composed of women (and men if they wish) from all walks of life who share a concern for the environment and its better management. Often the best insights come from women who do not perceive themselves as experts, but are living with environmental problems as part of their daily lives. The Forum structure permits the exchange of information among women and the raising of environmental awareness.

WorldWIDE News is issued six times a year. It contains: information from the Forums; profiles of women engaged in environment-related activities; and reports of success stories in environmental management. The newspaper discusses environmental problems, cites sources of environmental expertise and lists relevant events. Contributed articles and letters are welcomed and, to enhance networking, *WorldWIDE News* provides information on how contacts can be made. The World Resources Institute provides membership scholarships for women in non-industrialized nations so that they can receive free copies of the newspaper.

The first WorldWIDE directory was released in September 1987, listing women engaged in environmental activities by country and by area of interest and expertise. WorldWIDE is also compiling a list of women willing to speak on environmental subjects and welcomes information from women's organizations undertaking environmental activities. WorldWIDE monitors relevant legislation and policies, promotes regional and global activities to advance the participation of women in environmental matters and enables them to assist each other. It has, for example, recommended changes in the US foreign assistance legislation; connected women in different national parliaments to address the issue of women and environment; held meetings with the World Bank and environmental NGOs who have not adequately included women in their considerations; and provided forums on pesticides, chemicals, organic farming, water problems and the plight of indigenous peoples.

Y'S EYES

The Young Women's Christian Association (YWCA) is a worldwide network of thousands of women and women's organizations. Since 1955 the YWCA has been discussing energy-related issues, and since a 1979 Council meeting in Athens, it has addressed a number of environmental issues: pollution, the need to restore damaged areas, safety standards, and increasing pressures on the ecological system.

Environment and energy have become priority issues for World YWCA. They are closely related to the four other priorities of the organization: health, refugees, human rights and peace. At its 1984 Council meeting in Singapore, an Energy and Environment Cluster was organized. As a result, a YWCA network on energy and environment, called Y's EYEs, was established in 1985.

Y's EYEs's international network consists of national and local YWCA groups, of which more than 150 are active in the fields of energy and environment. One of the member groups' duties is to "be the all-seeing eyes. Look for what degrades the environment in your area, town, country. Look for polluters and expose them. 'Spy' on those who waste energy. And ... speak up" (Y's EYEs, January 1985: 2).

Two local YWCA's in Kenya are building shallow wells and laying gravity water pipes for village water supplies. In Uganda, a biogas demonstration centre has been established as a model in the garden of a YWCA member. Ghana's YWCA is working in the university with the Home Science Department, where students are housed in rural village settings and use technologies which make village life less difficult and more efficient. In Sierra Leone, YWCA members introduced improved cooking stoves to help save fuel and money. Their YWCA also started a study and action programme called "Health of Women as Affected by the Environment".

The YWCA in Trinidad and Tobago makes environment and energy issues part of its youth meetings, and organizes tree planting on World Environment Day. The Bolivian YWCA has planned a project on

"Preservation and Care of the Environment". In Zambia, appropriate technology documentation has been prepared by the YWCA and a library on appropriate technology is now in place. Valparaiso YWCA in Chile has a programme of social action in poor settlement communities of the city, focusing on appropriate technology. The Papua New Guinea YWCA is planning environmental programmes in schools and in the many squatter settlements around main towns. YWCA groups in the South Pacific, Canada, Greece, Lebanon and Britain are also active in the field of communication, information, coordination, action and lobbying.

CASE STUDIES

GREEN BELT MOVEMENT, KENYA

This programme of the National Council of Women of Kenya (NCWK) performs a double duty. Its central activity, tree planting, both reduces the effects of deforestation and provides a forum for women to be creative and effective leaders. By working with the Green Belt Movement, women can change their environment and make their own decisions. The Movement also involves the transfer of technology from experts to local people, enabling small-scale farmers to become agroforesters. Meetings related to tree planting can encompass discussions on wider issues, such as the relationships between food, population and energy. And Green Belt activities contribute to the raising of public awareness about environment and development.

The promotion of a positive image of and among women is one of the most important goals of the NCWK and its Green Belt Movement. Involving women as equal participants and developers of the green belts provides a good model of significant female achievement. Women are trained to plant and cultivate the seedlings, care for the trees and generate a source of income for themselves. Seedlings grown in project nurseries are sold to the Movement and then redistributed at no charge for new planting. The practical outcome of the work is also vital – communities become self-sufficient in fuelwood, and Kenyan women no longer spend hours each day in search of wood.

A further objective of the Movement is to involve the physically disabled and young school leavers; both have time to devote to the care of trees. It is hoped that the young people especially will become involved beyond the planting stage and want to remain in their communities, rather than migrate to the cities.

The most direct objective is environmental. The Green Belt Movement aims to create an understanding of the relationship between the environment and other concerns, such as food production and health. Education plays a key part in this: children are exposed to Green Belt projects at school and learn to appreciate the connections between forestry, soil conservation and their own need for wood.

How the project works

Groups or individuals who wish to participate must first prepare their land to meet the Movement's specifications. To foster high rates of tree survival, there is full discussion of the maintenance arrangements before planters receive any seedlings. An extensive follow-up programme also promotes success. Green Belt Rangers, often made up of the physically disabled, periodically check progress on the care of young trees and offer advice on problems. When a large green belt is planned, there is a ceremony, complete with important guests, to launch the planting to emphasize the significance of the action and heighten community awareness. These events allow the NCWK to meet local community leaders and establish critical ties between the women's group and other organizations.

Results

Thousands of trees have been planted and some 65 tree nurseries now exist. Not only has the Movement improved the local environment, providing trees for fuel, fodder, food and shelter, but it has involved large numbers of people, and responded to a multitude of needs. The Green Belt Movement has enjoyed local, national and international publicity. The image of Kenyan women has improved and public awareness of environmental issues has increased. In all, the Green Belt Movement confirms the essential connections between environmental conservation, the improvement of conditions for women and meeting the wider needs of society.

Source: *Case Studies from Africa*, 1987; NGO Liaison Service, UN.

AÇÃO DEMOCRÁTICA FEMININA GAÚCHA, BRAZIL

ADFG, or Friends of the Earth Brazil, is a voluntary, non-affiliated organization. When it was founded in 1964 as a women's organization, its main objective was to promote social change for equal opportunities. During its first ten years, ADFG emphasized educational work with girls and women, especially in poor urban areas. In 1983, men were permitted to become associates, although Board members are exclusively female and the emphasis on women's issues remains central to the organization.

Since 1974, another important issue has been added to ADFG's agenda: environmental protection and the promotion of sustainable development. At national and international levels, ADFG is fighting for a modification of

Brazil's prevailing agricultural system and for changes in development aid. ADFG campaigns and lobbies to have existing environmental protection laws respected and new ones created. It sees the mobilization of public opinion as a key element in forcing government to act. And, as we described in Chapter 2, ADFG also works at a grassroots level: in 1985 its Vacaria project began to oppose chemical-based agriculture.

Now, ADFG links national and international ecological organizations dealing with environmental conservation, peace and social justice. It also serves as an information centre for environmental organizations, for students, teachers, professional and governmental departments. ADFG is a member and representative of Friends of the Earth International, a founder member of the Pesticides Action Network (PAN) and its Latin American counterpart, and it coordinates PAN-Brazil. ADFG is also a member of the International Organization of Consumer Unions, the Environment Liaison Centre, the International Coalition on Energy and Development, the Coalition against Dangerous Exports, and the International Federation of Organic Agriculture Movements.

Source: María José Guazzelli, ADFG, Brazil.

A GUYANA WOMEN'S ORGANIZATION PROMOTES APPROPRIATE TECHNOLOGY

The Women's Revolutionary Socialist Movement (WRSM) had its beginnings in 1957 as the women's arm of the People's National Congress. WRSM has always promoted education and training for women, with particular emphasis on productive employment in support of the government's national self-sufficiency drive. Over the past fifteen years, it has diversified into economic and income-generating projects, promoting canteens, catering services and other food businesses as small commercial enterprises.

WRSM's projects include the production of textiles, ceramics, food, medical supplies, and arts and crafts and their associated marketing. An important underlying principle has been the maximization of the use of indigenous materials, particularly products derived from local crops which replace imported items. Another aspect is the development and dissemination of appropriate technology which reduces the drudgery of women's work and increases their efficiency in traditional activities, especially in the rural areas. WRSM operates, or is involved in the management of, fifteen projects which promote self-sufficiency in the meeting of basic needs.

The Movement's appropriate technology programme, originally funded by UNICEF, reaches out to satisfy community needs and is shared with eight Caribbean territories. As a result of the programme, a resource centre, the Appropriate Technology Centre, was established in Convention, a rural coastal area. The appropriate technology programme has fostered a number of different projects. Following a survey of the needs of women in three rural areas, a coconut mill was built. It is now run commercially by an engineering company. Stoves and ovens have been constructed by women for domestic and institutional purposes. Two day-care centres, one in Corriverton and one in Sand Creek, also arose from the appropriate technology programme. One is operated by WRSM personnel in cooperation with the municipality, the other is managed by the Ministry of Education for use as a nursery school.

The cultural advantages of promoting appropriate technology are particularly valuable in Guyana's multiracial society. The majority of the population is descended from Indians who were brought over in the nineteenth century as indentured servants. These people have used clay fires and chulas, and by adapting their methods, the appropriate technology workshop has been able to maintain their particular cultural pattern. Improvements carried out on the traditional fires by adding a chimney and an oven made it attractive to other members of the community because they were able to learn new cooking skills. In addition, the increased number of cooking places has enabled people to heat several pots and bake at the same time, so saving energy and money.

In another project, the Rice Van Shop, research is being conducted into ways of using indigenous flour, especially rice flour, so that imported wheat flour can be replaced. The unit was officially opened in Georgetown in 1984 and currently supplies a wide variety of rice-flour-based bread, pastry, cakes and other products to the city and outlying areas, while offering a lively catering service, including snackbars which are open every day and employ fifteen workers. The project is expanding into new outlets and is actively engaged in training women to produce its various products.

Source: Ovril Yaw, WRSM, Guyana.

ORGANIZATION OF RURAL ASSOCIATIONS FOR PROGRESS, ZIMBABWE

ORAP was formed in 1981 as a non-governmental organization to promote a new development strategy for the people of Matabeleland in Zimbabwe. Its aim was to respond to the plight of rural people who suffered during the struggle for independence. The environment in Matabeleland has been severely degraded. The number of people and grazing animals has increased on the poorest land (in colonial times, the best land was reserved for commercial agriculture), yet much of the poorer land is only marginally suitable for cultivation. The arable area has increased, causing overgrazing on the rest. Agricultural production has declined and there is an acute shortage of trees for fuel, thatching and fodder. Women are the primary victims of this environmental degradation, for they make up the bulk of the farmers here: their men have migrated to work in urban areas, in the mines and on the large agricultural estates.

ORAP began by helping village groups to form to discuss local needs and possible solutions. These groups then started to apply their labour and skills to practical work, helped – via ORAP – with materials and funds from donors. With the addition of more village groups, a federal structure of organizations was created, affiliated to ORAP, whose board is made up of elected representatives of the local groups. By June 1987, ORAP was serving 500 village groups (each with 50-100 local members) and had a staff of 42, two-thirds of them working at the village level.

The local groups have become a powerful force in improving conditions for women, and women play a prominent role in the whole programme. There is a hierarchy of groups for different activities. The primary "family unit" is created from up to 12 families who deliver direct and immediate benefits on such activities as house improvements, building latrines and water points. The group structure also supports these rural families by, for example, enabling them to pool their knowledge of child and health care.

Up to ten family collectives work together in "umbrella" groups on larger projects. These organize income-generating activities, including sewing (and school uniform making), basketry, building (and construction of building material), wood carving, poultry keeping, vegetable gardening, tree planting, baking, food storing, and improvements in water supply and sanitation. "Umbrella" groups have combined forces to create "associations" to undertake larger projects – grinding mills or dams, for example. Some associations have built "development centres" where neighbouring communities can meet, market their produce and run training workshops – in baking, blacksmithing and child care, for example. ORAP helps to organize these workshops by identifying people with skills and leadership qualities, and then training them in their special field. These "promoters" return to their own development centres, and build a workshop to pass on their new skills. Local people help voluntarily with the construction; ORAP provides funds and materials. These workshops encourage people to learn from, and experiment with, the latest thinking on appropriate technology and blend this with their traditional knowledge to tackle practical problems.

ORAP runs other training programmes for special groups, including women on its staff and in its membership. Women are helped, through discussion and practical research, to develop strategies for action. They have, for example, studied rural child-care patterns and identified the need for pre-school help while mothers are working. They have surveyed what families produce and can sell in the markets. ORAP also runs its own drought-relief programme, buying and transporting grain from food surplus areas and selling it, at cost price, in villages where there is a food deficit. With its experience of drought, ORAP gives a high priority to developing a comprehensive and sustainable food and water programme, based on a four-point strategy: the use of traditional seeds and fertilizers, the production of a variety of foods, improved food storage at village level, and better water management including local storage and irrigation schemes.

Source: ORAP and the International Labour Organization.

CHAPTER ELEVEN

The International Response

By 1975, international organizations, prompted by the UN Decade for Women, had begun to look at the role and position of women in development. How far have they come since then? And what exactly did the Decade for Women accomplish? This chapter reviews the Decade for Women and the policies of the major international environmental and development organizations, and assesses their role with regard to women and environment.

UN DECADE FOR WOMEN

In 1972, the year of the Stockholm Conference on the Human Environment, the General Assembly of the United Nations proclaimed 1975 International Women's Year. That year, the UN decided, would be devoted to intensified action to promote equality between men and women, to ensure the full integration of women in the total development effort and to increase women's contribution to the strengthening of world peace.

A World Conference on the International Women's Year was held in Mexico in 1975, adopted a World Action Plan to implement the objectives of the year, and proclaimed the period 1976-1985 the UN Decade for Women. Another world conference was held in Copenhagen in 1980, mid-way through the Decade, and an action plan for the next five years was agreed upon. In July 1985, the World Conference to Review and Appraise the Achievements of the UN Decade for Women: Equality, Development and Peace was held in Nairobi.

Representatives of 157 countries and 263 non-governmental organizations attended the Nairobi Conference, which had two principal goals: to review achievements of the Decade and discuss the obstacles to further successes, and to develop future strategies. *Forward-Looking Strategies for the Advancement of Women* formed the final document of the Conference.

According to Conference delegates, the overall achievement was the recognition of the essential role of women in development, both as beneficiaries and as contributors. The double burden of women – who must be income-earners as well as food producers and housekeepers – was recognized as an important obstacle to their equal participation with men in development. Delegates called upon governments and donors to devise development programmes which would take into account the drudgery of women and reduce their high unemployment and under-employment rates. The importance of increased access by women to land, credit, technology, markets, extension services and training was stressed.

But overall, the objectives of the Decade are far from being realized. Although it has generated information and a better understanding of the role and position of women in development, locally and globally, worldwide their situation has not been much improved; in some countries it has deteriorated.

Environmental concern
Representatives of several UN agencies, including UNEP, UNDP and WHO, focused on the relationship between women and the environment, and the problems environmental degradation poses for them. In the Nairobi Declaration, deep concern was expressed over the profound economic and social crisis that women and children were experiencing, especially in Africa, as a result of severe droughts, famine, external debt and the effects of the international economic situation. A key resolution generated at the Conference, and later adopted by the UN General Assembly, urged: international agencies to provide more information on women's role in conservation; that governments take environmental factors into consideration in their development projects; that social as well as economic criteria be applied to development; and that sustainable development be promoted.

The main outcome of the Nairobi Conference, however, was the document *Forward-Looking Strategies for the Advancement of Women*, which was accepted by all 157 countries and later adopted, without a vote, by the 40th Session of the UN General Assembly in resolution 40-108 on 13 December 1985. It dealt with a range of issues: equality, development (including agriculture, science and technology, housing, energy, environment, education, employment, health, industry, trade and commercial services, communication and social services), peace, international and regional cooperation, and special areas of concern (such as women in drought areas, the urban poor, elderly and young women, women deprived of their traditional means of livelihood, sole supporters of the family, refugee and indigenous women). But what signalled a change was the fact that, for the first time in the history of the Women's Decade, the issue of the environment – as it related to women – was taken up. The relevant sections, adopted by consensus, are:

- Deprivation of traditional means of livelihood is most often a result of environmental degradation resulting from such natural and man-made disasters as droughts, floods, hurricanes, erosion, desertification, deforestation and inappropriate land use. Such conditions have already pushed great numbers of poor women into marginal environments where critically low levels of water supplies, shortages of fuel, over-utilization of grazing and arable lands and population density have deprived them of their livelihood. Most seriously affected are the women in drought-afflicted arid and semi-arid areas and in urban slums and squatter settlements. These women need options for alternative means of livelihood. Women must have the same opportunity as men to participate in the wage-earning labour force in such programs as irrigation and tree planting and in other programs needed to upgrade urban and rural environments. Urgent steps need to be taken to strengthen the machinery for international economic cooperation in the exploration of water resources and the control of desertification and other environmental disasters (paragraph 224).
- Efforts to improve sanitary conditions, including drinking water supplies, in all communities should be strengthened, especially in urban slums and squatter settlements and in rural areas, with due regard to relevant environmental factors. These efforts should be extended to include improvements of the home and the work environment and should be effected with the participation of women at all levels in the planning and implementation process (paragraph 225).
- Awareness by individual women and all types of women's organizations of environmental issues and the capacity of women and men to manage their environment and sustain productive resources should be enhanced. All sources of information and dissemination should be mobilized to increase the self-help potential of women in conserving and improving their environment. National and international emphasis on ecosystem management and the control of environmental degradation should be strengthened and women should be recognized as active and equal participants in this process (paragraph 226).
- The environmental impact of policies, programs and projects on women's health and activities, including their sources of employment and income, should be assessed and the negative effects eliminated (paragraph 227).

Implementation of the objectives of the Decade on equality, development and peace, and of the resolutions in *Forward-Looking Strategies*, will be monitored over the years 1986 to 2000, both at national and international levels.

Central in this review and appraisal will be the UN Commission on the Status of Women, established by the UN Economic and Social Council in June 1946. The Commission's mandate is to prepare recommendations and reports on the promotion of women's rights in political, economic, social and educational fields. In 1967, the Commission adopted the Declaration on the Elimination of Discrimination against Women, which was approved by the General Assembly in 1979 and came into force in 1981. The Commission is now setting its sights on the year 2000 as a target date in the effort to achieve a just society in which dignity, opportunity and power are not the monopoly of one sex.

Other specialized United Nations agencies which focus particularly on the role and position of women are the UN Development Fund for Women (UNIFEM), the International Research and Training Institute for the Advancement of Women (INSTRAW), and the Branch for the Advancement of Women at the Centre for Social Development and Humanitarian Affairs.

But, like all international movements for change, it will take time for the recommendations of the Women's Decade to become a reality and for the benefits to be realized at the local level. Meanwhile, other international organizations are beginning to recognize the crucial part that women play in environmental management.

INTERNATIONAL UNION FOR CONSERVATION OF NATURE AND NATURAL RESOURCES

IUCN is an independent, international organization with over 500 members, including states, government agencies and NGOs in 116 countries. IUCN's purpose is to ensure, as far as possible, that the process of socio-economic development throughout the world is sustainable and that the biosphere is managed for the overall benefit of present and future humankind.

IUCN was founded in 1948 as a result of concern throughout the international scientific community about the extent to which economic development and the growth of the human population were causing significant adverse and often irreversible effects on the environment. Its focus then was protecting individual species and habitats. The link between conservation and development has been a recurrent theme within IUCN over the last thirty years, but recently it has assumed increasing importance. It was the central argument of the *World Conservation Strategy* (WCS) prepared by IUCN, UNEP and the World Wildlife Fund in collaboration with the FAO and UNESCO. The *Strategy* stressed the critical importance of sustainable development, and directed world attention to the increasingly dangerous stresses being placed on the earth's biological systems. The goal of the *World Conservation Strategy* was to integrate conservation and development to ensure the survival and well-being of all people through the maintenance of ecosystems, preservation of the biological diversity of species, and the sustainable use of both.

IUCN's three-year working programme, the "Conservation Programme for Sustainable Development", provides the basic framework within which its activities, including field and support projects, are planned and executed. Its Operations Division currently looks after ninety field projects throughout the world.

Essential components of the IUCN network are its six commissions covering ecology; environmental education; sustainable development; national parks and protected areas; policy, law and administration; and species survival. These commissions are made up of more than 1,200 experts in 120 countries and over 2,000 correspondents and consultants, who constitute the intellectual resources of IUCN and keep it abreast of conservation problems and new solutions.

IUCN operates the Conservation Monitoring Centre in the United Kingdom, the IUCN Environmental Law Centre in Bonn, Federal Republic of Germany, and the Conservation for Development Centre (CDC) at the Secretariat in Gland, Switzerland. The terms of reference of CDC are to obtain tangible sustainable benefits for humankind, especially the poorest communities, through the application of practical conservation principles. CDC, backed by a worldwide registry of consultants, helps countries to ensure that their resources are used to provide sustainable benefits. Its specialities are the preparation and implementation of environmental management plans and the development of the human resources and institutions needed to manage natural resources effectively. For example, CDC provided the IUCN's Secretariat with research material for the *Sahel Report*, a long-term strategy for environmental rehabilitation of the Sahel (1986).

Women and IUCN
At IUCN's 16th General Assembly in Madrid in 1984, a recommenda-

tion was put forward to guarantee better representation of women and women's issues in its work. As the then president, Mohammed Kassas, said:

> We are yet to achieve true participation of women in the work of the Union. Few women have been invited to lead discussions at this General Assembly and I would like us to look into prospects and means for bringing women all over the world into the movement. Let us all take this mission to heart.

In 1987, only nine of the 30 professional staff members of IUCN's Secretariat were women. The representation of women in IUCN's expert commissions is poor. In the Commission on Sustainable Development the number of female experts is 18 compared with 135 men – little more than 13 per cent – and this is one of the commissions in which women are most highly represented. In the Conservation for Development Centre's consultant register only seven per cent of consultants are women. This situation is no different from that in many other international organizations and institutes.

The representation of professional women in development organizations is important. But more crucial is that the donor and policy-making organizations have a proper focus on the role and position of women. Some IUCN documents, such as the Union's *Bulletin*, do this. The IUCN *Sahel Report*, under the heading "Promoting New Styles of Development at the National and International Level", says:

> New perceptions of the role of women are crucial for a re-evaluation of the contribution of major groups who have been historically disadvantaged. These include especially women and children who constitute the majority of the rural population as a result of socio-economic forces, particularly rural-urban migration.
>
> Care should be taken to avoid gender segregation that will be counter-productive. Women's actual economic and social activities, including farming of food crops, rather than preconceived ideas about their feminine role, should be targeted for particular assistance (IUCN, 1986).

National conservation strategies

The World Conservation Strategy (WCS) was meant to be used by individual countries as a framework for developing their own national conservation strategies. By March 1987, these national strategies had been finalized for nine countries, and twenty were in the pipeline.

The implementation of these strategies has been slow and not altogether satisfactory. Few countries have put their plans into action, although, as the case study of Zambia's Luangwa Project shows, there are exceptions. In the strategies prepared to date, no specific reference has been made to the role of women in the use and management of natural resources, nor does it seem that women's groups and organizations have been much involved in their implementation. In only two countries, Pakistan and Malaysia, have women been closely involved. And in Botswana, where the government's national conservation strategy seminars attracted large numbers of women farmers, women's opinions were not solicited or encouraged (K.P. Walker, Forestry Association Botswana, 1986).

In the spring of 1986, an international conference on "Conservation and Development: Implementing the World Conservation Strategy" was held in Ottawa, Canada, to review the worldwide implementation of the WCS and to see what policy changes were needed. A caucus on "Women, Environment and Sustainable Development" proposed that a supplement to the WCS be prepared on women and environment. The following recommendations were adopted by the conference:

Women, Environment and Sustainable Development

The specific place of women in promoting sustainable development is increasingly appreciated. However, there is still an overall lack of serious consideration of the role of women, their contribution and potential in relation to environment and development, and there has been a failure to allocate sufficient resources to ensure their inclusion and integration.

Rec. 1. Environmental organisations including IUCN, development agencies, and donors should actively develop, support and implement policies, strategies and plans of action which will integrate women fully into sustainable development and natural resources conservation and management, both as part of their own process and of the full and proper implementation of the World Conservation Strategy and other conservation strategies.

Rec. 2. IUCN and other international bodies should allocate sufficient resources to support assessments of the impacts on women of development and conservation projects.

Women, Environment and Sustainable Development

The specific role of women, and particularly indigenous women, in relation to environment and development is insufficiently recognized and supported, and is not considered in the World Conservation Strategy.

Rec. 3. The sponsors of the World Conservation Strategy should promote a supplement on Women, Environment and Sustainable Development to be drafted by a special task force, including members from the caucus held at the Ottawa Conference on Conservation and Development.

A Working Group on Women, Environment and Sustainable Development has now been convened under the auspices of IUCN and is reviewing the *World Conservation Strategy*, and considering ways in which IUCN might adapt its own programme to incorporate women's issues.

ENVIRONMENT LIAISON CENTRE

The ELC was established in Nairobi, Kenya, in 1974 by NGOs concerned about environment and development. It has over 230 non-governmental member organizations in more than 65 countries, and maintains contacts with over 7,000 environment and development groups. ELC's major objectives are:

- to strengthen NGOs working in the field of environment and sustainable development, particularly in the Third World, through liaison, provision of information, training and financial assistance;
- to build links between environmental NGOs around the world and between environment and development NGOs; and
- to facilitate NGO contributions to and support for the United Nations Environment Programme (UNEP), the United Nations Centre for Human Settlements (UNCHS) and, where appropriate, other international organizations.

While ELC's main concerns are environmental – afforestation, energy, sustainable agriculture, for example – it has devoted some attention to women. At its February 1985 global meeting on environment and development, a women's caucus emerged which, among other things, proposed that chair persons at NGO meetings take up the practice of alternately calling on women and men. (Needless to say, the recommendation has not been adopted.) More important was a series of ELC-organized workshops on women and environment held at

Forum '85 in Nairobi, where participants from all over the Third World presented case studies on women's involvement in forests, energy, sustainable agriculture and water management.

The main results of the workshops were a final statement, submitted by the ELC to the UN Conference on the Decade for Women, and a series of recommendations dealing with women and political change, development, community participation, and women's participation. But the real meaning of the Nairobi gathering was best summed up by Mary Ann Eriksen, ELC adviser, in her introduction to the workshop report:

> These women speak directly to us and we must listen carefully to what they say, for their world is our world and their problems ours. Finally, of course, we must do more than listen. We must lead the way out of the dreadful situation in which the world finds itself. Each of us must work in our own land, each in our own job and in our own way. We must work with existing organizations and institutions and leaders when we can, we must redirect them when necessary. We must find new methods and establish new organizations when the old fail us. We must lead. And we must join together in a concerted effort to make the world a better place, everywhere, for people to live. We must prove that Nairobi was only the beginning (Mary Ann Eriksen, in ELC, 1986:3).

With these words in mind ELC is now working on a follow-up project on women, environment and development, which will be integrated into the overall ELC programme of activities.

UNITED NATIONS ENVIRONMENT PROGRAMME (UNEP)

The United Nations Environment Programme (UNEP), headquartered in Nairobi, was established in 1972 as an outcome of the UN Conference on the Human Environment. This agency, the first to be established in a developing country, has regional and liaison offices in Bangkok, (Manama) Bahrain, Geneva, Mexico City, New York City and Washington, D.C. With a small professional staff of some 180 people, UNEP serves as catalyst and coordinator on environmental issues within the United Nations system, among governments, NGOs and regional groups. It is run by a Governing Council from 58 member states and has centres addressing issues such as regional seas, desertification, terrestrial ecosystems, chemicals and pesticides, global resource assessment and monitoring, environmental law, and environmental education. UNEP is also engaged in projects connecting environment to health, energy and technology. It operates a 'clearing house' by which donors can be matched with specific environmental projects presented to UNEP by individual governments.

In 1982, at UNEP's tenth Governing Council, the organization was directed to expand its efforts by reaching out beyond its traditional partners to engage the interest and action of new groups, including industry, parliamentarians, women, health workers, religious leaders and young people.

UNEP and women

UNEP has been a leader in connecting the issues of women and environment. In 1980, in preparation for the United Nations Decade for Women Conference in Copenhagen, UNEP issued a series of brochures which identified the relationships between women, natural resources and environmental systems. These booklets, widely distributed in Copenhagen, helped the United Nations Secretariat for the Nairobi Women's Conference to prepare key resolutions in *Forward-Looking Strategies for the Advancement of Women.*

In 1985, the Executive Director of UNEP, Dr Mostafa K. Tolba, determined that UNEP would undertake several initiatives to encourage the participation of women in managing the environment in conjunction with UNEP's outreach programme and the UN World Conference to Review and Appraise the Achievements of the United Nations Decade for Women.

UNEP's first initiative was to provide funding for ELC to organize the workshop on women and the environment as part of the NGO Forum. The next initiative was to establish a UNEP Committee of Convenors for the Nairobi Women's Conference, composed of senior women interested in environmental issues who would be serving on their nation's delegations and other women in leadership positions on environmental matters. This committee of twenty distinguished women met for the first time in Nairobi just before the Women's Conference. They came prepared to: encourage adoption of key paragraphs in the *Forward-Looking Strategy* document; introduce a resolution calling on women to take a leadership role in achieving sustainable development through environmental and natural resource management; and host an evening seminar for delegations and NGO leaders to present a framework for connecting women and environmental issues to development and peace, two of the conference's key themes.

The rationale for these two activities was the belief that, regardless of their origin or culture, women are taught to manage prescribed resources from the time they are children, whether it is so much food for so many mouths, so much space for so many to sleep, so many potatoes from so many fields, or so much money from selling in the market place or the husband's pay cheque. This tradition is invaluable preparation for women as they move into playing their part in sustainable development.

In his address, "An Alliance with Nature: Women and the Earth's Traditions", Dr Tolba told delegates:

> I look to you, women from all walks of life, to join us in defining and, most critically, redirecting the course of development to prevent further environmental catastrophes. The burden of environmental degradation and crises has always fallen and is still falling on women, especially in the developing countries. Women who are in positions of influence have a special duty to represent those at the sharp end of the environmental crises. You, as more than half of the human race, must mobilize. You, who often suffer first and are consulted last, you, who must live with the consequences of decisions without having a forum to voice your objections; you must join us. If there must be war, let it be against environmental contamination, nuclear contamination, chemical contamination, against the bankruptcy of soil and water systems; against the driving of people away from their lands as environmental refugees. If there must be war, let it be against those who assault people and other forms of life by profiteering at the expense of nature's capacity to support life. If there must be war, let the weapons be your healing hands, the hands of the world's women in defence of the environment. Let your call to battle be a song for the Earth.

UNEP also prepared a background document outlining the connections between the environment and other conference topics such as employment, health, education, food, water and agriculture, industry, trade and commercial services, science and technology, and energy. In addition, a series of fact sheets on specific environmental problems, such as desertification, water, deforestation, ozone and health were prepared in English, Arabic, Spanish and French, and distributed at the conference. They suggested action that individuals and institutions could take to address these problems, and are available from UNEP.

Over 800 non-governmental and delegate leaders attended the UNEP seminar. Five of the senior women gave presentations on environmental issues in Africa, Latin America and the Caribbean, industrial nations, the Middle East, and Asia in the aftermath of Bhopal. The seminar helped to focus women's thinking on the relevance of environmental issues to development policy and peace. UNEP now maintains a women's network listing the participants' location and areas of special interest. In addition, UNEP now has a committee of Senior Women Advisers on Sustainable Development and a women's officer.

In April 1987, the Senior Women were invited to consider how they might regionalize their activities and work cooperatively with UNEP's regional offices. Meeting in Nairobi in June 1987, and in response to an invitation from the Minister of Natural Resources and Tourism of Zimbabwe, the Honourable Mrs Victoria Chitepo (a Committee member), the Senior Women decided to initiate a regionalized effort by supporting an African Women's Assembly in October 1988 in Zimbabwe to assist the African Ministers' Cairo Plan of Action.

The Cairo Plan was developed by African Environmental Ministers in 1986, at the instigation of UNEP. Its aim was to mobilize Africans to reverse the environmental degradation of the continent and to achieve energy and food self-sufficiency at the village level. In response to the plan, eighteen nations in Africa have each identified three villages to be the focus for achieving these goals. The Senior Women Advisers, through the African Women's Assembly, propose to assist this process by involving women, especially from the designated countries and villages, to develop "recipe books" of specific actions which address one of the four areas of concern identified by the African Ministers: forests, fresh-water lakes and rivers, arid and semi-arid lands and coasts, at the village level.

UNEP's Senior Women Advisers are sharing information about their own work and helping each other to organize resources and action ranging from field projects to influencing government policies. They are active in areas which involve reforestation, pesticide and sanitation projects, environmental legislation and education, environment-related health issues, air, land, water, noise pollution, and women in environmental organizations.

In 1986, the UN Secretariat for the Advancement of Women designated UNEP as the lead agency for women and the environment in the UN system. In early 1987 the final report of the Women's Decade was amended, calling on women to promote sustainable development. This reflected a strongly held position by UNEP's Senior Women Advisers during the Nairobi Conference – that the Women's Decade initiative should call for "Equality, Sustainable Development and Peace".

In June 1987, at UNEP's 14th Governing Council, a resolution was passed deciding that UNEP's 1988 *State of the Environment Report* would emphasize women and the environment. This resolution is but another positive reflection of the creative partnerships which have evolved among UNEP, its Senior Women Advisers on Sustainable Development, member governments and the United Nations' system.

UNIFEM

UNIFEM is a fund which supports the development initiative of poor women in the Third World who are outside the reach of mainstream development agencies. Created in 1976 by the UN General Assembly, it was first called the Voluntary Fund for the UN Decade for Women. Renamed the UN Development Fund for Women (UNIFEM) in 1985, it works in association with the United Nations Development Programme (UNDP). UNIFEM has two primary objectives:

- to provide direct financial and technical support to women involved in cooperative activities, food production, fuel and water supply, health services, small businesses, management, and planning; and
- to ensure that the needs of both women and men receive consideration when large-scale assistance is given to developing countries – through involvement in programming and project design, monitoring and evaluation.

UNIFEM depends on voluntary contributions from governments, organizations and individuals to support its activities. Its slogan is: "Progress for women is progress for all."

DEVELOPMENT ASSISTANCE

Many government and other donor agencies officially recognize women's issues and environmental factors in their bilateral and multilateral aid programmes. They issue statements, present papers at international meetings and adopt project guidelines. But, so far, much of the commitment is only on paper and, as the following brief analysis shows, few of these organizations link women and the environment in the implementation of their policymaking.

The Australian Development Assistance Bureau, for example, ensures that all its project appraisals and evaluations include considerations of women, environment and sustainable development, although the bureau has not so far initiated projects which link these issues. There is a Women in Development Unit, a steering committee which monitors government policy and a special fund to support women's interests.

DANIDA, the Danish International Development Agency, is working with Danish women's organizations on an action plan for development assistance to women. Environmental issues are considered most important in formulating strategies for social and economic development in recipient countries and DANIDA is planning to increase its environmental expertise. There is a working group on the role of women in development policies, and every aid mission has a staff member responsible for women's issues. In a growing number of DANIDA's environmental rehabilitation projects – concerned with soil conservation, afforestation and the reduction of sand drift – women have been identified as the target group, or are playing a major part in project preparation and implementation. The agency funds small as well as large-scale projects and in these, the implications for women are always assessed.

In the Netherlands, the Ministry of Foreign Affairs houses a unit on International Women's Affairs which is focused on development assistance. The Dutch aid programme evaluates its schemes for their effects on women, supports special women's projects and works with women's organizations. In 1987 the Minister of Development Assistance presented an action programme on women and development, which aimed to promote the active participation of women. The expertise in the field of women and development will be expanded, and in the whole project cycle the interests and needs of women will be taken into account. An independent Commission on Ecology and Development Assistance, reporting in 1986, recommended procedures for project design, implementation and evaluation to incorporate environmental issues and suggested the preparation of ecological profiles for those countries in which Dutch aid is concentrated. The Commission called for more restoration projects, more support for environmental NGOs and the expansion of ecological staff and expertise in the Dutch Ministry and embassies. While there is still room for more coordination of women and environment, the importance of this linkage is officially recognized in Netherlands policies and in Dutch support for projects such as the Energy and Rural Women's Work Programme and the Women, Environment and Sustainable Development project on which this book is based.

In the United Kingdom, the Overseas Development Administration (ODA) is committed, at least on paper, to the notion that women are important agents of development. There are no special projects and no earmarked funds for women's development, but the policy is to integrate consideration of women's needs into the work of all the departments of ODA. There is an environmental adviser and two social development advisers who share (along with their other tasks) the giving of advice on women's issues. But this advice is not automatically sought or given for every ODA-funded project: a checklist on women's roles and needs is applied only if a project is judged as liable to have some impact on them, and in practice, it seems that few projects are so judged (Mazza, 1987). Women's issues do feature, in a limited way, in the ODA's training and research programmes. There are no women in the most senior decision-making positions.

The US Agency for International Development (USAID) has a special Office of Women in Development which reviews projects. The agency has moved away from women-specific schemes, believing that integrated projects are a more effective approach and that women's issues should be included in initial project designs. A portion of USAID's funds are directed towards increasing women's incomes.

The West German Aid Agency (BundesMinisterium Wirtschaftliche Zusammenarbeit), like most others, takes a sectoral approach to environmental issues. Both conservation and women are important themes in the aid programme and women have been identified as a special target group since the beginning of the Women's Decade. Projects, especially those on rural development and drinking water supply, are evaluated for their impact upon women's lives.

Aid for women – still an afterthought

Clearly, some development assistance agencies are more active than others in their consideration of women's development needs and in their linkage of the related issues of women and environment. But most are still better on paper than in practice. A recent critique of the British aid programme concludes: "aid can do more harm than good – especially to women" and suggests that some of the conventional beliefs of aid policy are damaging women's interests (Mazza, 1987). Despite the US approach, it is, for example, by no means the case that *any* increase in household incomes is good for women: the benefits of improved male incomes rarely "trickle down" to wives. Meanwhile, women's roles as subsistence farmers, providers and full-time carers remain invisible to most development planners. Some projects have profoundly damaged women's lives. The Bura cotton project in Kenya, backed by ODA and the World Bank, allowed cash-croppers to improve their production on irrigated land. But women subsistence farmers, with no access to irrigated plots, had to walk further each day to cultivate poorer land. Mazza claims the project has brought waterborne disease and severe malnutrition, especially to children. In a Zambian Integrated Rural Development Project, there is increased malnutrition among the children of women who now have less time to care for them as they work harder on their husbands' cash-crops. Karnataka's Social Forestry project in India, designed to help poor farmers, has likewise brought few benefits for poor women who must now walk further to collect their fuelwood.

Where development assistance is directed at women, it often presupposes their role as passive recipients of aid, rather than as active agents of change. They and their families are seen as a drain upon local economies – not as potential contributors. In general, NGO (rather than government) aid agencies such as OXFAM and Christian Aid take a more positive view of women in development: they are increasing the number of small-scale enabling projects in sustainable farming which can benefit the poorest. It is true that national governments often support these projects, but over the last decade, aid budgets have dwindled. The British aid programme, for example, had fallen from 0.52 per cent of national income in 1979, to 0.34 per cent by 1985 – an amount far short of the UN target of 0.7 per cent of GNP. And the focus of much aid spending has shifted away from projects designed to help the poorest and towards large-scale prestige projects which tie development assistance to exports and the promotion of foreign policy objectives.

Multilateral aid agencies

The World Bank, Regional Development Banks, the European Development Fund (financed by EEC members), and UN agencies such as UNICEF and WHO have all made some adjustment to their policies to incorporate women's issues and the environment. The World Bank has adopted guidelines which check its projects for their impact on both. A new approach to project definition and management is being tried out in a number of recipient countries where Bank action plans and programmes will aim to assist women. The Bank has researched several "women in development" studies and is compiling a bibliography on "women in resource management".

Following the 1979 World Conference on Agrarian Reform and Rural Development, the Food and Agriculture Organization (FAO) prepared operational guidelines for the integration of women in rural development. Similar guidelines are used to check that projects assisted under FAO's World Food Programme do not damage women's interests. More recently, FAO has begun a number of schemes in which women are the main target group, including, for example, an action project for disadvantaged women in Kenya's arid lands, and FAO now gives increased support for environmental rehabilitation programmes in afforestation and soil conservation.

In 1983, the Development Assistance Committee (DAC) of the Organization for Economic Cooperation and Development (OECD) adopted some guiding principles for supporting the role of women in development to which the member states of OECD are committed. The guidelines argue for the incorporation of women's interests in all aspects of development policy and call for specific projects to benefit women by improving their income generation and their access to land, credit and training. The DAC guidelines emphasize the need for good local consultation arrangements on aid projects, where "the advice of local women's groups is likely to be particularly valuable" (OECD, 1983). DAC is presently discussing with OECD member states a proposal to collect statistics which will show the extent to which women benefit from the aid projects funded by them.

In all, there are encouraging signs of a move among most aid agencies to respond to the messages of the Women's Decade, although the practice still falls short of the ideals set out in the *Forward-Looking Strategies* document of the Nairobi Conference. Projects which aid the poorest are starved of funds; there are too few safeguards of women's interests in major aid projects; there is too little participation by women and their organizations in the development process; and training for women is inadequate. Environmentally, many projects fail. For procedures to change, so must attitudes. The need now is not just to reduce the damage that many aid projects still do, but to shift aid spending towards investing in women as a major resource for improving the environment and thereby the economy and welfare of communities in the Third World.

CASE STUDY

LUANGWA INTEGRATED RESOURCE DEVELOPMENT PROJECT, ZAMBIA

The Luangwa Integrated Resource Development Project, encompassing an area of about 15,000 square kilometres in the eastern, central and northern provinces of Zambia (including the South Luangwa National Park), is a result of the National Conservation Strategy for Zambia. The project was formally begun in 1986 and supported by the Zambian government, NORAD, WWF-International and WWF-USA, and IUCN. The main objective of the project is to improve the standard of living of the people of the area through sustainable use of the full range of natural resources available to them, including land, forests, fisheries, water and wildlife. Integration of conservation with development, the promotion of greater self-reliance, decentralization and diversification of activity in the framework of the National Conservation Strategy are central goals of the project.

A women's programme is included. Women are disadvantaged in African rural society since they occupy a subordinate position in male-headed households. In female-headed households, they are subordinate to other households. In rural Zambia, the proportion of female-headed households averages around 35 per cent of the total. Women are nutritionally, culturally and economically less affluent than Zambian men, the problem being particularly severe in the female-headed households. The objective of the women's programme in the Luangwa Project is to provide an extension service which improves the nutrition and health of women in the project area, with the intention of breaking the poverty cycle. The need for male involvement in the programme is recognized.

For the second phase of the Luangwa Project, 1988-1992, it is proposed to increase the staff to eight field workers and a coordinator for the women's programme. This group will initiate:

- agricultural extension directed at women's clubs, women and female-headed households, and emphasizing protein crops such as soybeans;
- provision of credit facilities for agricultural inputs, domestic industries and small processing plants;
- poultry farming;
- better crop storage;
- the improvement of rural water supplies through the rehabilitation of existing wells and the installation of new ones and mobile training courses for women, with particular emphasis on nutrition, hygiene and health.

Source: Luangwa Integrated Resource Development Project, Proposals for the Phase 2 Programme, LIRDP Project Document no. 3, June 1987.

CHAPTER TWELVE
Working Together For The Future

*It all depends on your philosophy of life. If you dream about an egalitarian
and just society it cannot be created by men alone. You cannot ignore 50
per cent of the population* (Shekhar Pathah, India, 1986).

Around the globe, women are shaping the environment – and caring for
it. For most of them in the Third World, especially the poorest, a
healthy environment is fundamental to their survival. The environment
is not just a backdrop to their activities, but impinges on all aspects of
their lives; it conditions their livelihood and the welfare of their families.
Women as providers and carers are wholly dependent upon the
renewability of natural systems to provide for their basic needs of food,
water and shelter. They rely upon natural biomass sources for food,
whether this is hunted like fish, gathered, or grown as crops. The
environment provides fodder for their animals and fuel for their fires,
whether these use wood, crop wastes or dung. Women look to natural
systems for fertilizers, building materials, medicines and the ingredients
of many of their income-earning enterprises, especially in food-
processing and craft work.

Earlier chapters, adopting the messages of the World Conservation
Strategy and the World Commission on Environment and Development, have argued that conserving the environment is a fundamental
ingredient of sustainable development – the only kind of development
that will benefit women (IUCN, 1980; WCED, 1987). In essence, this
means managing the environment to maintain its life-support systems,
protect its biological diversity and ensure that its species and ecosystems
are used in ways that are sustainable.

In all these tasks women are vital, for in many areas of the world, they
are the main environmental managers. This is particularly true in
Africa, where women grow most of the food. "Women are the
backbone of Africa," says Harrison, and here and elsewhere they are
traditionally responsible for a great deal of conservation activity – they
protect soils, trees and water sources, they carefully select the seedcorn

damage and they pioneer ways of reclaiming land severely degraded by bad management.

WHAT IS HAPPENING AND WHY

But the close and symbiotic relationship that women have with the natural environment – built up over generations – is breaking down.

Earlier chapters have shown that environmental degradation is accelerating for many reasons. Foremost among these is the emphasis of much development and investment in the Third World on cash-cropping. This has displaced traditional practices, forcing subsistence farmers into marginal areas with consequent deforestation and desertification. In 40 years, Africa has moved from a position of self-sufficiency in food to deficit: Kenya alone, with large tracts of land under cash crops, now imports more than six kilos of food a year for each person. Even on family plots, where men have title to the land, cash crops may displace food grown by women for their families and preclude them from traditional conservation practices. They spend long – and unpaid – hours tending the cash-crops for which the men receive payment. And in spite of their large share of the workload, women are excluded from the crop planning and many other farm management decisions. As Haleh Afsar says: "It is men who own the world's resources even when it is women who make them productive" (Afsar, 1987).

Resettlement schemes which have transferred poor farmers to poor land have often brought similar results. Women's traditional rights to cultivate land have frequently been removed by such schemes and they have lost the incentive to conserve (Rogers, 1983). The migration of men away from their families to seek jobs in the cities has further increased women's work burden and reduced the time they can devote to environmental care. In many countries of the South, population growth exacerbates all these pressures on natural resources, especially around cities, contributing to a spiral of degradation. In western Sudan, for example, the pressure to produce more food for more people has caused farmers to abandon crop rotation. This has brought soil erosion, reduced yields, stimulated more intensive cropping and so induced more erosion.

The poverty of many Third World families has meant that more and more must rely on the "free" goods they can collect from their surroundings. As supplies of fuelwood and fodder are exhausted, especially around cities and refugee camps, the resilience of the environment to recover is lost. Meanwhile women must forage further and further from home, reducing the time and energy they have not only for environmental repair but for other tasks to help their families. Collecting wood and water can consume 500 calories a day from a total intake that, for women, is often less than 1500 calories, even when they are pregnant (Taylor, 1985).

The tragedy is that remedial programmes of environmental rehabilitation often ignore women's needs and fail to build upon their capacity to conserve. Both development and environment projects are to blame here. Some social forestry schemes, for example, as we examined in Chapter 4, have displaced women. Environmental conservation projects – those, for example, funded by the Worldwide Fund for Nature (previously World Wildlife Fund) – still concentrate mainly on the protection of species and habitats which are valued in the North rather than considering the survival of whole ecosystems, including their human residents (WWF, 1987).

Moreover, the training and support programmes (including credit) which might enable small farmers to manage their land well are most often directed at men. In many countries women are denied title to land, cannot enter into contracts and cannot raise loans. Technologies, too, are transferred mainly to men: it is they rather than women who are reached by agricultural extension programmes and who are taught how to install and maintain water pumps or biogas plants. Grinding mills and stoves are used by women but designed and built by men.

As this book has shown, environmental destruction hits women hardest. They already suffer from their lowly, unequal position in the household and in the community and their inferior status compared with men. They carry a heavier work burden, and perform more time-consuming tasks on the land and in the home. They have a poorer diet, shabby working conditions, bad health and inadequate health care. They have little or no control over cash or land. When young women migrate to the city for some improvement in their living conditions, many end up in the lowest paid, most menial occupations in overcrowded, polluted factories producing electronic components and textiles – the "sweatshops of the sun" (Mitter, 1986). One in three of the world's women are illiterate. Worldwide, their position seems to get worse.

Women as the key agents of change

Yet women are the ones who could make a major contribution to environmental rehabilitation. First, they have the knowledge and the skills of natural resource management that can be built upon. They know, for example, much better than imported "experts" which trees make the best fuelwood, which dry fast and burn well, which retain moisture in the soil and which ones give the best foliage for fodder or fertilizer.

Secondly, women have a remarkable ability to work together. The evidence from many women's groups is that women can share their skills and resources to take effective action. Some of the most successful examples of sustainable development build upon women's initiatives – in the Chipko movement of northern India, in the Kenyan Green Belts and in the Baldia sanitation project in Pakistan. Many women are experienced in cooperative working, something that male-dominated governments and multinational companies have yet to learn (Mukhopadhyay, 1984).

Thirdly, in caring for children, women have a powerful influence over changing attitudes to the environment. They can turn the short-term expediency of remedial action into a lasting habit of environmental protection.

Finally, perhaps most important of all, it is likely that restoring women's capacity to care for the environment will be associated with improvements in their independence and their status: there is a major convergence of interest between environmentally sound and sustainable development, and the development of women.

Agnes Aidoo comments:

> Existing development policies that marginalize women contribute to the crisis in Africa but also women are only marginally involved in the search for and implementation of solutions to environmental degradation. If this trend is allowed to continue one important avenue will be lost for redirecting African development (Aidoo, 1985).

The need now is to move on from the perception of women as victims (or even as villains) of the environmental crisis and instead to assist them as major agents of rehabilitation.

The global response

The last decade has seen increasing recognition of the scale of the environmental crisis and the failings of conventional development. There has been growing concern about the plight of women. All three issues – environment, development, women – are now more firmly placed on the international agenda and, as the last chapter showed, they are beginning to permeate the thinking and the action of national and international donors. Many NGO welfare agencies, working directly with local communities, already translate women's needs into practical action. There are good examples from many smaller NGOs operating in the South and especially from women's own organizations (Harrison, 1987).

The Nairobi Women's Conference generated useful discussion and many papers, especially in Forum '85. A women's caucus at the Ottawa Conference on the World Conservation Strategy kept the issues of women and environment alive and a number of global networks are continuing to promote awareness. But is this enough?

WHAT IS NEEDED

In most Third World countries, women are among the poorest of the poor. And poverty, it is now accepted, is a major cause of environmental degradation (WCED, 1987). It is unlikely that either the environment or the living conditions of women will be substantially improved until the major causes of poverty in the South are tackled. According to the World Commission on Environment and Development, that means addressing not just the local symptoms of distress, but the underlying reasons for the appalling and self-defeating mismanagement of resources, both human and natural. Foremost is the debt crisis which forces Third World countries so dangerously to exploit their resource base. The Commission calls for a major effort to improve food security through supporting small food producers with appropriate price and other incentives.

Population growth, low and fluctuating commodity prices, dwindling and inappropriate aid programmes and the escalation of arms spending are other problems which national governments must tackle with more than rhetoric. The Commission sees institutional change as a major way forward and argues not just for more effective environmental protection laws and agencies, but for development organizations of all kinds to be directly accountable for the consequences of their actions (WCED, 1987).

How, then, can women improve their position? The *Forward-Looking Strategies* document of the Nairobi Conference is explicit about what is needed: "Effective development requires the full integration of women in the development process as both agents and beneficiaries. Development agencies should take full cognizance of women as a development resource" (Para. 308, FLS, 1985). In 372 recommendations, the *Strategies* set out ways for women to achieve equality in health, education, training, technology and information, land, income and credit. The recommendations echo a long-held philosophy:

Women must see that there can be no liberation for them and no solution to the ecological crisis within a society whose fundamental model of relationships continues to be one of domination.... How do we change the self-concept of a society from the drives towards possession, conquest and accumulation to the values of reciprocity and acceptance of mutual limitations? (Rosemary Radford Ruether, World YWCA Day, 1982).

The Nairobi call is for a massive transfer of power to women at all levels to enable them to take more control over their own development. It is tempting to argue that all the lobbying effort should be devoted to this end. But, realistically, many of the recommendations will not be implemented soon, if at all. There are too many vested interests in resisting them and formidable obstacles to change, not least in the political, cultural and religious circumstances in which many women must live.

So what, in the meanwhile, can be done to build on the good work that has come out of the Women's Decade?

Summarizing the practical recommendations on the theme of women and the environment that have emerged recently (including those of the *Forward-Looking Strategies*, of the Environment Liaison Centre's Plan of Action, from UNEP's Senior Women Advisers and from the Ottawa Conference), the following seem to be the priorities for action:

- Improving women's capacity to conserve and benefit from sustainable development at the local level;
- Improving their access to training; and
- Continuing awareness-raising and advocacy.

Improving women's capacity

It is by starting with the priorities of the poorer, and enabling them to gain the livelihoods they want and need, that both they and sustainable development can best be served (Chambers, 1987).

The case studies in this and other books show that already there are successful practical examples of sustainable development which point the way (Harrison, 1987; Timberlake, 1987; Chambers, 1987). They suggest that a crucial ingredient for success is that projects should be locally designed and locally managed. There are many popular organizations which already engage in conservation work when they are given the resources to do so. The Naam groups in Burkina Faso, for instance, relying on local knowledge and low-cost local tools and materials, have brought about a highly successful style of village development, controlled by villagers, which includes water conservation projects, stove improvements, woodlots and vegetable gardens. The role of government and donors has been to provide technical and some material support to augment local efforts.

Women's groups (although they sometimes exclude the poorest) have been especially effective in mobilizing support for conserving activities. There are more than 6,000 women's groups in Africa alone, and many of these, set up as savings clubs and craft cooperatives, have shown themselves to be willing and flexible enough to organize environmental tasks which respond to local needs.

The more successful projects integrate several aspects of environmental conservation and development. Tree planting offers soil protection, fuelwood and fodder, and, with fruit and nut trees, even food. New water points can be linked to improvements in health care; biogas developments can be associated with vegetable gardening and nutritional advice. For women who manage homes and farms and families, this kind of integration is a natural way of working.

But to take on conservation work – where returns are rarely immediate – women must have legal and secure rights to cultivate land. Development projects may need to help women negotiate for these rights. Wherever possible, projects should link environmental improvements to new income-earning opportunities, for these increase women's independence, status and negotiating power in getting credit, marketing advice and other supports. And income-earning is not only about women making crafts for sale (where the market is often poor and volatile) but about trading in food and cash crops, charcoal and other products that can be sustainably made from local materials. Many women's groups have shown that these enterprises can be combined with subsistence food production. Access to family-planning information and services is another essential ingredient. Unless women have control over their own fertility and reproductive health, they are unlikely to be able to contribute substantially to practices that sustain development – or raise their own status.

There are demonstration projects with all these elements, but they are far too few. The lessons of pioneering schemes need now to be replicated in many more places. Too often projects of this kind are seen as optional extras, to be cut when time and budgets run short on large-scale, conventional developments. More governments, donors and UN agencies need to recognize their long-term – and not just pioneering – value, and to fund grassroots groups to organize them or arrange for welfare NGOs, acting as intermediaries, to do so. "The approach," says Harrison, is "basic and unglamorous ... but ... we are dealing here with foundations and first steps. Only if the steps are sound, and well-directed, is there any hope of further progress" (Harrison, 1987).

Access to training

The success of primary health care has shown how effective it can be to "train the trainers", and create, at local level, trained people who can pass on skills and information. The same can be done for conservation. Individual women like Ione Halley (Chapter 5) can be highly effective "animateurs", persuading people to adopt new environmental practices. More women like Ione should have access to training in leadership as well as in conservation techniques.

But for this to happen, women must be a target group for training, and arrangements must be made, as they were in one Zimbabwean experiment, for women to be able to leave their farms and families to attend special workshops. Studies by the International Labour Organization suggest that working with women separately is often necessary to ensure they are reached by training programmes. Further, local training works best where it is augmented by government extension programmes (where more women must be recruited), and by communications on radio and in the press. It helps where there are networks of school and youth clubs promoting similar messages, as is the practice in the wildlife clubs of Kenya and Zambia.

In the past, the emphasis has been on Northern experts training those in the South (again, mainly men). Now the focus must shift towards much more South-to-South training, with women being assisted to pass on their knowledge, not only from village to village, but between countries. Such schemes do exist. One example is the volunteer technical assistance programmes of the African Training and Research Centre for Women in Addis Ababa. Another is the technical assistance programme of AFOTEC in Senegal (case study on pages 126–8). Northern donors could help here, perhaps with a special fund, from which they could pay for travel and other means for grassroots groups to exchange personnel and information, not just about technical matters but about how best to organize themselves. These groups need funds to produce case studies of their projects, to plan and run training workshops, distribute training materials, form local networks for "primary environmental care" and pioneer ways of spreading conservation messages using modern audio-visual methods as well as the traditional media of stories and songs, dance and drama. Much information is already transferred in these ways but it has yet to be well documented and shared.

Awareness and advocacy

The momentum of the Women's Decade continues to bring changes in the way national and international development agencies *say* they will operate. Progress so far seems rather slower among environmental agencies, such as IUCN, although the preparation and implementation of national conservation strategies which IUCN is now promoting could offer a practical means of highlighting the issue of women and environment and of encouraging women's groups to participate in conservation measures. There is now, in some organizations, more effective screening of projects (throughout the project cycle) for their impact on women's lives. Pressure must now be put upon all those governments who have not done so to spell out how the new procedures are being implemented.

Women are still not well represented among senior decision-makers in development institutions. One of the surprises of the report of the World Commission on Environment and Development (considering its chairman, Gro Harlem Brundtland, heads a cabinet of whom half are women) is that there is so little about women in the document – and indeed only three of the 19 commissioners were women. Brundtland says that "women are used to thinking in the long term". Yet at the top, as at the grassroots, they are not listened to.

Global networks, like those of WorldWIDE and IUCN, are helping women to be heard, but these are mainly the professionals. There are too few lines of communication between international and local organizations; this limits women's ability to influence the ways in which development is planned and implemented. The need is for national or regional focal points, with a funded secretariat, to form alliances of environmental and welfare NGOs, research and educational institutions, and women's groups which can argue against misguided projects and press for sustainable development action. Where their dignity and confidence have been restored, women have shown they can act decisively: they **do** have something to say.

REFERENCES

INTRODUCTION

Aidoo, Agnes Akosua (1985). *Women and Environmental Rehabilitation.* ATRCW/ECA.

IUCN (1980). *World Conservation Strategy: Living Resource Conservation for Sustainable Development.* IUCN/UNEP/WWF.

Rocheleau, Dianne (1985). *Women, Environment and Development: a question of priorities for sustainable rural development.* Background paper, Global Meeting on Environment and Development for NGOs, Nairobi, 4-8 February 1985.

Senghor, Diana (1985). "Feminism and Environmentalism: In a candle light", in *Ecoforum*, Vol. 10:2 (April 1985).

WCED (1987). *Our Common Future.* Oxford University Press.

World Resources Institute and International Institute for Environment and Development (1986). *World Resources 1986.* New York: Basic Books.

CHAPTER 1. WHY WOMEN?

Fresco, Louise (1985). "Vrouwen en Landbouwontwikkeling; de stand van zaken 10 jaar na Mexico", in *Geen Oplossingen zonder vrouwen*. NCO (Nationale Commissie voor Voorlichting in Bewustwording Ontwikkelingssamenwerking) Platformbijeenkomst Plattelandsontwikkeling en de Positie van de Vrouw, Arnhem, 24 April 1985.

ILO (1986). *Energy and Rural Women's Work.* Vol. II. *Papers of a Preparatory Meeting on Energy and Rural Women's work, Geneva, 21-25 October 1985.* Geneva: ILO.

Taylor, Debbie *et al.* (1985). *Women: A World Report.* London: Methuen and New Internationalist.

CHAPTER 2. LAND: WOMEN AT THE CENTRE OF THE FOOD CRISIS

Agarwal, Bina (1986). *Cold Hearts and Barren Slopes: The Woodfuel Crisis in the Third World*. New Delhi: Allied Publishers Private Limited and Institute of Economic Growth.

Ahmed, Iftikhar (1985). *Technology and Rural Women: Conceptual and Empirical Issues*. London: Allen & Unwin.

Arizipe, Lourdes, and Josefina Aranda (1986). "Women Workers in the Strawberry Agribusiness in Mexico", in Eleanor Leacock and Helen Safa (eds), *Women's Work: Development and the Division of Labor by Gender*. South Hadley, Massachusetts: Bergin & Garvey, Inc.

Bamba, Nonny (1985). "Ivory Coast: Living with Diminishing Forests", in *Women and the Environmental Crisis*. Report of the Proceedings of the Workshops on Women, Environment and Development, Nairobi, 10-20 July 1985. Nairobi: Environment Liaison Centre.

Bandyopadhyay, J., and Vandana Shiva (1986). *Drought Development and Desertification*. Brief Report on the 2-day Seminar on the Control of Drought Desertification and Famine, held at India International Centre, New Delhi, 17-18 May 1986.

Barry, Jessica (1986). "Oxen: a Women's Best Friend?", in *Earthscan Bulletin*, April 1986.

Brakel, Manus van (1986). "Het is belangrijk uit te gaan van onze eigen kennis en technieken. Boeren en Boerinnen uit de Derde Wereld (2)", in *Milieudefensie*, Vol. 15 (1986), no. 1.

Braun, Arnelle (1984). "The First Person you See is a Buffalo", in *Ceres*, Special issue on 'Food, Agriculture and Women'. Rome: FAO.

Bull, David (1982). *A Growing Problem: Pesticides and the Third World*. Oxford: Oxfam.

Chaney, Elsa (1985). *Subsistence Projects for Rural Women*. Kellogg Institute.

Creevey, Lucy (ed) (1986). *Women Farmers in Africa: Rural Development in Mali and the Sahel*. Syracuse NY: Syracuse University Press.

Dankelman, Irene (1985). "Vrouwen, Milieu en Ontwikkeling", in *Nieuwsbrief Milieu and Ontwikkeling*, Vol. 3 (4).

Davies, Miranda (1983). *Third World – Second Sex: Women's Struggles and National Liberation*. London: Zed Press.

Eviota, Elizabeth U. (1986). "The Articulation of Gender and Class in the Philippines", in Eleanor Leacock and Helen Safa (eds), *Women's Work: Development and the Division of Labor by Gender*. South Hadley, Massachusetts: Bergin & Garvey, Inc.

FAO (1981). *Agriculture: Toward 2000*. Rome: FAO.

FAO (1986). "Women are Farmers too", News Release, February 1986.

Feminist Theory, State Policy and Rural Women in Latin America. Kellogg Institute Working Paper, no. 49, December 1985, Notre Dame.

Foster, Theodora (1986). *A Common Future for Women and Men (and All Living Creatures): a submission to the World Commission on Environment and Development*. EDPRA Consulting Inc., Ottawa, Canada, 31 March 1986.

Fresco, Louise (1985). "Vrouwen en Landbouwontwikkeling; de stand van zaken 10 jaar na Mexico", in *Geen Oplossingen zonder Vrouwen*. NCO Platformbijeenkomst Plattelandsontwikkeling en de Positie van de Vrouw, Arnhem, 24 April 1985.

Guazzelli, María José (1985). "Southern Brazil: Breaking with an Imposed Dependence", in *Women and the Environmental Crisis*. Report of the Proceedings of the Workshops on Women, Environment and Development, Nairobi, 10-20 July 1985. Nairobi: Environment Liaison Centre.

Gubbels, P.A., and A. Iddi (1986). *Women Farmers: Cultivation and utilization of soybeans among West African women through family health animation efforts. Case Study*. Oklahoma City: World Neighbors.

Huston, Perdita (1985). *Third World Women Speak Out*. Asian Women's Research and Action Network, Davao City, Philippines.

ICDA (1985). "Traditional Varieties: the Testimony of a Farmer", in *ICDA News Special Report: Seeds*, July 1985.

IFDP, Rural Women's Research Team (1980). *Tough Row to Hoe: Women in Nicaragua's Agricultural Cooperatives*. San Francisco: IFDP.

ILO INSTRAW (1985). *Women in Economic Activity: a Global Statistical Survey (1950-2000)*. Statistical Publication no. 10, INSTRAW.

ISIS (1983). *Women in Development: A Resource Guide for Organisation and Action*. ISIS Women's International Information and Communication Service. Geneva: ISIS.

Jiggins, Janice (1984). "Food Production and the Sexual Division of Labour; Policy and Reality", unpublished paper.

Lappé, Frances Moore, and Joseph Collins (1978). *Food First*. New York: Ballantine.

Lappé, Frances Moore, and Joseph Collins (1986). *World Hunger: Twelve Myths*. New York: Grove Press; London: Earthscan.

Leon de Leal, Magdalena (1985). "State Rural Development: Colombia", in *Feminist Theory, State Policy and Rural Women in Latin America*, Kellogg Institute Working Paper, no. 49. University of Notre Dame, December 1985.

MAG (1986). Proyecto Determinación de Residueos de Pesticidas Clorados en Leche Materna. Ministério de Agricultura y Canaderia y Consejo Nacional de Ciencia y Tecnología. Quito, December 1986.

Muntemba, S. (1985) (ed). *Rural Development and Women – Lessons from the Field*. Geneva: ILO.

Nash, June, and Helen Safa *et al.* (1986). *Women and Change in Latin America*. South Hadley, Massachusetts: Bergin and Garvey Inc.

Netherlands IUCN Committee (1986). *The Netherlands and the World Ecology: Towards a National Conservation Strategy in and by the Netherlands, 1986-*

1990. Amsterdam: Netherlands National Committee for IUCN.

Non-Governmental Liaison Service (1987). *Case Studies from Africa: Towards Food Security*. New York: United Nations.

Nyoni, Sithembiso (1985). "Africa's Food Crisis: Price of ignoring Village Women?", in *Women and the Environmental Crisis*. Report of the Proceedings of the Workshops on Women, Environment and Development, Nairobi, 10-20 July 1985. Nairobi: Environment Liaison Centre.

Pearce, Jenny (1986). *Promised Land*. London: Latin American Bureau.

Poldermans, Caroline (1985). "Rol van de Vrouw in Voedselproduktie wordt onvoldoende Onderkend", in *Aspecten*, Vol. 18 (6).

Sachs, Karin (1982). "An Overview of Women and Power in Africa", in Jean F. O'Barr (ed), *Perspectives on Power*. Durham, N. Carolina: Duke University Center for International Studies.

Shiva, Vandana (1985). "India: the Abundance Myth of the Green Revolution", in *Women and the Environmental Crisis*. Report of the Proceedings of the Workshops on Women, Environment and Development, Nairobi, 10-20 July 1985. Nairobi: Environment Liaison Centre.

Shiva, Vandana (1985-2). "Where has all the water gone? The case of water and feminism in India", in *Women and the Environmental Crisis*. Report of the Proceedings of the Workshops on Women, Environment and Development, Nairobi, 10-20 July 1985. Nairobi: Environment Liaison Centre.

Sinha, Rhadha (1984). *Landlessness: a growing problem*. Rome: FAO.

Spears, J.S. (1978). *Wood as an Energy Source: the Situation in the Developing World*. Paper presented at the 103rd Annual Meeting of the American Forestry Association. Washington D.C.: The World Bank.

Taylor, Debbie *et al.*(1985). *Women: A World Report*. London: Methuen and New Internationalist.

UNEP (1980). *Women, Environment and Food*. Nairobi: UNEP.

Weiss, Ruth (1986). *Die Frauen von Zimbabwe*. München Frauenbuchverlag, Weisman Verlag.

Whitehead, Ann (1985). "The Green Revolution and Women's Work in the Third World", in Wendy Fulkner and Erik Arnold (eds), *Smothered by Invention: Technology in Women's Lives*. London: Pluto Press.

Williams, Paula J. (1984). *The Women of Koundougou*. Hanover, USA: Institute of Current World Affairs.

World Council of Churches (1986). "Women Ending Hunger", in: *Women in a Changing World*. Geneva: WCC.

World Resources Institute and International Institute for Environment and Development (1986). *World Resources 1986*. New York: Basic Books.

CHAPTER 3. THE INVISIBLE WATER MANAGERS

Chauhan, Sumi Krishna, *et al.* (1983). *Who puts the Water in the Taps? Community participation in Third World drinking water, sanitation and health.* London: Earthscan.

CIDA (1985). *Women, Water and Sanitation.*

IDWSSD (1984). *Insights from Field Practice: How women have been and could be involved in water supply and sanitation at the community level.* Inter-Agency Task Force on Women and Water of the IDWSSD Steering Committee for Cooperative Action. New York: IDWSSD.

IDWSSD (1985). *Strategies for Enhancing Women's Participation in Water Supply and Sanitation Activities. Recommendations of the Inter-Agency Task Force on Women and Water, of the IDWSSD Steering Committee for Cooperative Action. New York: IDWSSD.*

KWAHO (1984). *Training Women in Maintenance and Use of Simple Water Supply Systems. Report of the KWALE Workshop, 22-26 October 1984. Nairobi: Kenya Water for Health Organization.*

Nicholson, S.E. (1982). *The Sahel: A climatic perspective.* Club du Sahel. Sahel D (82): 187.

Soon, Young Yoon (1983). *The Women's Dam: The Mossi of Upper Volta.* Regional Office of WHO of Asia and Pacific Region.

Timberlake, Lloyd (1985). *Africa in Crisis.* London: Earthscan.

UNDP (1985). *Is There a Better Way? Promotion and Support for Women's participation in the International Drinking Water Supply and Sanitation Decade. A project of UNDP, 1981-1990.* New York: UNDP.

UNEP (1982). *Women, Environment and Water.* Nairobi: UNEP.

UNICEF and INSTRAW (1985). *Women and the International Drinking Water Supply and Sanitation Decade.* Paper submitted to The World Conference to Review and Appraise the Achievements of the UN Decade for Women. Nairobi, 15-25 July 1985.

WCED (1987). *Our Common Future.* Oxford University Press.

WHO (1981). *Decade Dossier (International Drinking Water Supply and Sanitation Decade 1981-1990).* Geneva: World Health Organization.

Wijk-Sijbesma, Christine van (1985). *Participation of Women in Water Supply and Sanitation: Roles and Realities.* International Reference Centre for Community Water Supply and Sanitation. Technical Paper 22. The Hague: IRC.

Wijkman, Anders, and Lloyd Timberlake (1984). *Natural Disasters: Acts of God or Acts of Man?* London: IIED/Earthscan.

World Resources Institute and International Institute for Environment and Development (1986). *World Resources 1986.* New York: Basic Books.

CHAPTER 4. WOMEN AND FORESTS

Agarwal, Anil, and Anita Anand (1982). "Ask the Women who do the Work", in *New Scientist*, 4 November 1982.

Aloo, Theresa (1985). "Forestry and the Untrained Kenyan Women", in *Women and the Environmental Crisis*. Report of the Proceedings of the Workshops on Women, Environment and Development, Nairobi, 10-20 July 1985. Nairobi: Environment Liaison Centre.

Bahuguna, Sunderlal (1984). "Women's Non-Violent Power in the Chipko Movement", in Kishwar, M. (ed), *In Search of Answers: Indian Women's Voices from Manushi*. London: Zed Press.

Bamba, Nonny (1985). "Ivory Coast: Living with Diminishing Forests", in *Women and the Environmental Crisis*. Report of the Proceedings of the Workshops on Women, Environment and Development, Nairobi, 10-20 July 1985. Nairobi: Environment Liaison Centre.

Bhatty, Zarina (1984). "Women in Forestry: India", in *Tigerpaper*, Vol. 11 (3).

Cecelski, Elizabeth (1985). *The Rural Energy Crisis, Women's Work and Basic Needs: Perspectives and Approaches to Action*. ILO Rural Employment Policy Research Programme Technical Co-operation Report. Geneva: ILO.

Centre for Science and Environment (1985). *State of India's Environment: Second Citizen's Report, 1984-85*. Delhi: Centre for Science and Environment.

Chavangi, Noel A., Rutger J. Engelhard and Valerie Jones (1985). *Culture as the Basis for Implementing Self-Sustaining Woodfuel Development Programs*. Nairobi: The Beijer Institute.

Cruz, Cerenella A. (1984). "Women in Forestry: The Philippines", in *Tigerpaper*, Vol. 11 (3).

Draper, Patricia (1975). "!Kung Women: Contrasts in Sexual Egalitarianism in Foraging and Sedentary Contexts", in Rayna R. Reiter (ed), *Toward an Anthropology of Women*. New York and London: Monthly Review Press.

FAO (1983). *Rural Women, Forest Outputs and Forestry Projects*. Discussion Draft. Rome: FAO.

Fortmann, Louise, and Dianne Rocheleau (1985). *Women and Agroforestry; Four Myths and Three Case Studies*. ICRAF Reprint no. 19. Nairobi: ICRAF.

Foster, Theodora (1986). *A Common Future for Women and Men (and All Living Creatures). A submission to the World Commission on Environment and Development*. EDPRA Consulting Inc., Ottawa, Canada.

French, David (1986). *Agroforestry for Food, Fuel and Income: Four project approaches focusing on women's work*. Edited and abridged paper prepared for the ILO, Geneva.

Hoskins, Marilyn (1979). *Women in Forestry for Local Community Development*. Washington D.C.: USAID, Office of Women in Development.

McCall Skutch, Margaret (1986). "Participation of Women in Social Forestry

Programmes: problems and solutions", in *BOS Newsletter*, Vol. 5 (1).

Musumba, Brazille (1985). *KENGO's Role in Energy and Environment Development in Kenya*. Nairobi: KENGO.

Obel, Elizabeth (1985). *The Impact of Afforestation Programmes through Women's Groups in Kenya*. A paper submitted for the Earthscan fellowship grant for NGO writers. Nairobi: KENGO.

Paris, Thelma R. (1986). *Women in a Crop-Livestock Farming Systems Project in Santa Barbara, Pangasinan, Philippines*. Paper prepared for the Conference on Gender Issues in Farming Systems Research and Extension, University of Florida, Gainesville, Florida, 26 February-1 March 1986.

Ramanankasina, Estelle (1986). "African Women Brace for Advancing Desert", in *Women and the Environmental Crisis*. Report of the Workshops on Women, Environment and Development, Nairobi, 10-20 July 1985. Nairobi: Environment Liaison Centre.

Reiter, Rayna R. (ed) (1975). *Toward an Anthropology of Women*. New York and London: Monthly Review Press.

Sheth, Malini Chand (1985). "Indian Women in Defence of Forests", in *Women and the Environmental Crisis*. Report of the Workshops on Women, Environment and Development, Nairobi, 10-20 July 1985. Nairobi: Environment Liaison Centre.

Shiva, Vandana, J. Bahdyopadhyay and N.D. Jayal (1985). "Afforestation in India: Problems and Strategies", in *Ambio*, Vol. 14 (6).

Spears, J.S. (1978). *Wood as an Energy Source: the situation in the developing world*. Paper presented at the 103rd Annual Meeting of the American Forestry Association. Washington D.C.: The World Bank.

Thrupp, Lori-Ann (1984). "Women, Wood and Work. In Kenya and Beyond", in *Unasylva*, December 1984.

Ummaya, Pandurang and, J. Bandyopadhyay (1983). "The Trans-Himalayan Chipko Footmarch", in *The Ecologist*, Vol. 13 (5).

Williams, Paula J. (1984). *The Women of Koundougou*. Hanover, USA: Institute of Current World Affairs.

Williams, Paula J. (1985-1). Women and Forestry. Paper for IX World Forestry Congress, Mexico, 1-10 July 1985.

Williams, Paula J. (1985-2). *Women's Participation in Forestry Activities in Burkina Faso*. Hanover, USA: Institute of Current World Affairs.

World Resources Institute and International Institute for Environment and Development (1986). *World Resources 1986*. New York: Basic Books.

CHAPTER 5. WOMEN'S ENERGY CRISIS

Agarwal, Bina (1986). *Cold Hearts and Barren Slopes: The woodfuel crisis in the Third World*. New Delhi: Allied Publishers Private Limited and Institute of Economic Growth.

Alcantara, Elsa (1985). "The Domestic Energy Crisis, Women's Work and Family Welfare in three Ecological Areas of Peru", in Volume II of papers of a preparatory meeting on Energy and Rural Women's Work, Geneva, 21-25 October 1985. Geneva: ILO.

Ardayfio-Schandorf, E. (1986). The *Rural Energy Crisis in Ghana: its implications for women's work and household survival*. ILO Working Paper, World Employment Programme. Geneva: ILO.

Ardayfio-Schandorf, E. (1984). *Energy and Rural Women's Work in Ghana. Preliminary Results*. Report to the ILO, University of Ghana, Legon.

Bajracharya, Deepak (1983). *Rural Energy Planning in the Developing Countries of Asia*. Background Paper for Presentation at FAO/ESCAP Seminar on Rural Energy Planning. 9-29 April 1983, Beijing, People's Republic of China.

Barnard, G., and L. Kristoferson (1985). *Agricultural Residues as Fuel in the Third World*. London: Earthscan.

Bart, F. (1980). "Le Paysan Rwandais et L'Energie", in *L'Energie dans les Communautes Rurales des Pays du Tiers Monde*. Centre d'Etudes de Géographie Tropicale, Bordeaux, 5-10 May 1980.

Caceres, Armando (1986). *Rural Household Improvement by Integrated Bioenergy Systems in Guatemala*. Presented at the International Workshop on the Rural Energy Crisis, Women's Work and Basic Needs, The Hague, 21-24 April 1986.

Cecelski, Elizabeth (1985-1). *The Rural Energy Crisis, Women's Work and Basic Needs: Perspectives and Approaches to Action*. ILO Rural Employment Policy Research Programme. Technical Co-operation Report. Geneva: ILO.

Cecelski, Elizabeth (1985-2). *Energy and Rural Women's Work. Issues for Discussion*. Paper prepared for the Preparatory Meeting on Energy and Rural Women's Work, ILO, Geneva, 21-25 October 1985.

Cecelski, Elizabeth (1986). *Energy and Rural Women's Work: Crisis, Response and Policy Alternatives*. Background paper prepared for the International Workshop on the Rural Energy Crisis, Women's Work and Basic Needs. 21-24 April 1986, The Hague, The Netherlands. Geneva: ILO.

Centre for Science and Environment (1985). *State of India's Environment: Second Citizen's Report, 1984-85*. Delhi: Centre for Science and Environment.

Courier (1986). "The Woodfuel Crisis", in *The Courier*, no. 95, January-February 1986.

Cuanalo, M.E. de (1983). *The Use of Firewood as Domestic Fuel in Mexico and the Patterns of Change to Alternatives*. Institute of Agricultural Economics,

University of Oxford, August 1983.

Eckholm, Erik, Gerald Foley, Geoffrey Barnard and Lloyd Timberlake (1984). *Fuelwood: the energy crisis that won't go away.* London: Earthscan.

Elmendorf, M. (1980). *Women, Water and Waste: Beyond Access.* Washington DC: Equity Policy Center.

Elnur, Awatif Mahmoud (1985). "Renewable Options for the Sudanese Women", in *Women and the Environment Crisis.* A Report of the Proceedings of the Workshops on Women, Environment and Development, Nairobi, 10-20 July 1985. Nairobi: Environment Liaison Centre.

Fleuret, Patrick C., and Anne Fleuret (1978). "Fuelwood Use in a Peasant Community: a Tanzanian Case Study", in *Journal of Development Areas,* April 1978.

Foley, Gerald, and Patricia Moss (1983). *Improved Cooking Stoves in Developing Countries.* Earthscan Technical Report, no. 2. London: Earthscan.

Foster, Theodora (1986). *A Common Future for Women and Men (and All Living Creatures): a submission to the World Commission on Environment and Development.* EDPRA Consulting Inc., Ottawa, Canada.

Gattegno, I., and J. Muchnik (1983). *Rapport de Mission Haute Volta: la fabrication du dolo et l'économie de bois de feu.* ALTERSIAL/ENSIA, Paris.

Gebreselassie, Alasebu (1985-1). *Household Fuel Energy Planning with the Participation of Women in Ethiopia.* A paper submitted to the Preparatory Meeting on Energy and Rural Women's Work, 21-25 October 1985. Geneva: ILO.

Gebreselassie, Alasebu (1985-2). "An Experience in the Improvement of Employment Conditions of Rural Women in Ethiopia: The Case of the Golgotta Settlement Horticultural Initiative and Dana I Settlement Goat-Raising Project for Women", in Shimwaayi Muntemba (ed) *Rural Development and Women: Lessons from the Field.* Vol. II. Geneva: ILO, 1985.

Gupta, Vrinda (1985). "Women in the Lead: India and Improved Cookstoves", in *Women and the Environmental Crisis.* A report of the Proceedings of the Workshop on Women, Environment and Development, Nairobi, 10-20 July 1985. Nairobi: Environment Liaison Centre.

Hayes, P. (1982). Rural energy consumption in Somalia. Draft report of the Ministry of Mineral and Water Resources, Mogadishu, Somalia.

Hoskins, Marilyn W. (1979). *Women in Forestry for Local Community Development.* Washington D.C.: USAID, Office of Women in Development.

ILO (1966). *Maximum permissible weight to be carried by one worker.* Report to the International Labour Conference, Fifty-first session, Geneva. Geneva: ILO.

Khamati, Beatrice (1985). "Women and Cookstove Development", unpublished paper, KENGO. Nairobi: KENGO.

Koenig, D. (1984). *Energy Use in the Regions of Nioro and Bougonni.* Draft paper for USAID, Mali Renewable Energy Project. Washington, DC: USAID.

Manibog, F. (1984). "Improved Cooking Stoves in Developing Countries. Problems and Opportunities", in *Annual Review of Energy*, Vol. 9.

Musumba, Brazille (1985). *KENGO's Role in Energy and Environment Development in Kenya*. Nairobi: KENGO.

Muzira, Mary Tereza (1985). *Women and the Woodfuel Crisis in Uganda*. Resource Paper for UN Decade for Women, NGO Forum '85. Jinja: JEEP.

Nagbrahman, D., and Shreekant Sambrani (1983). "Women's Drudgery in Firewood Collection", in *Economic and Political Weekly*, 1-8 January 1983.

Puerbo, Hasan (1985). *Rural Women and Social Structures in Change: A Case Study of Women's Work and Energy in West Java, Indonesia*. Indonesian Rural Women's Work and Energy Project Team, Centre for Environmental Studies, Bandung Institute for Technology and Centre for Development Studies, Bogor Agricultural Institute.

Sarin, M. (1983). *Improved Woodstove Dissemination by Village Women: the case of the Nada Chulha*. Paper presented at the International Workshop on Woodstove Dissemination, Wolfheze, Holland, 31 October-11 November 1983.

Schuler, S. (1981). The Women of Baragaon. Vol. II, Part 5 of *The Status of Women in Nepal*. Kathmandu, Nepal: CEDA.

Skar, S.L. (1982). *Fuel Availability, Nutrition and Women's Work in Highland Peru*. ILO, Geneva, 1982, mimeographed World Employment Programme Research Working Paper (restricted).

Skutch, Margaret M. (1985). *The Success of Mixed Motives: stove and forestry programmes in Gujarat*. Enschede: Technology and Development Group.

Smith, K.R., A.L. Agarwal and R.M. Dave (1983). *Air Pollution and Rural Fuels: implications for policy and research*. WP-83-2. Honolulu, Hawaii: Resource Systems Institute, East-West Center.

Spears, J.S. (1978). *Wood as an Energy Source: the Situation in the Developing World*. Paper presented at the 103rd Annual Meeting of the American Forestry Association. Washington, D.C.: World Bank.

Srinavasan, V. (1984). *Mahila Narajagran Samiti (Women's Reawakening Association)*. Draft Report to the ILO/DANIDA project on successful income-generating projects for women.

Tinker, Irene (1984). *The Real Rural Energy Crisis: Women's Time*. Washington, DC: Equity Policy Center.

Ummaya, Pandurang, and J. Bandyopadhyay (1983). "The Trans-Himalayan Chipko Footmarch", in *The Ecologist*, Vol. 13 (5).

WHO (1984). *Biomass Fuel Combustion and Health*. Geneva: World Health Organization.

Williams, Paula J. (1984). *Women and Cookstoves*. Hanover, USA: Institute of Current World Affairs.

Wisanti Utama, Elly (1985). "Indonesia: an NGO approach to facing a women's crisis", in *Women and the Environmental Crisis*. A Report of the Proceedings of the Workshops on Women, Environment and Development, Nairobi, 10-20 July 1985. Nairobi: Environment Liaison Centre.

World Resources Institute and The International Institute for Environment and Development (1986). *World Resources 1986*. New York: Basic Books.

CHAPTER 6. HUMAN SETTLEMENTS

Agarwal, Anil (1983). *Mud, Mud*. London: Earthscan.

Butalia, Subhadra (1985). "Bhopal: Women bear brunt of environmental disaster", in *Women and the Environmental Crisis*. Report of the Proceedings of the Workshops on Women, Environment and Development, Nairobi, 10-20 July 1985. Nairobi: Environment Liaison Centre.

Cecelski, Elizabeth (1985). *The Rural Energy Crisis, Women's Work and Basic Needs: Perspectives and Approaches to Action*. Rural Employment Policy Research Programme, Technical Co-operation Report. Geneva: ILO.

Celik, Aliye (1985). "Human Settlements: a Critical Factor", in *Women and the Environmental Crisis*. Report of the Proceedings of the Workshops on Women, Environment and Development, Nairobi, 10-20 July 1985. Nairobi: Environment Liaison Centre.

Davidson, J. (1985). "Human Settlements: Building a New Resourcefulness", in *Habitat International*, Vol. 9 (3/4).

Hardoy, J.E., and D. Satterthwaite (1984). "Third World Cities and the Environment of Poverty", in *Geoforum* 15.

Hardoy J.E., and D. Satterthwaite (1986-1). "Urban Change in the Third World", in *Habitat International*, Vol. 10 (3).

Hardoy, J.E. and D. Satterthwaite (1986-2) "Shelter, Infrastructure and Services in Third World Cities", in *Habitat International*, Vol. 10 (3).

Hosken, Frau (1987). "Women, Urbanisation and Shelter", in *Development Forum*, May 1987.

Kudat, A. (1986). *Women and Shelter*. Nairobi: UNCHS.

Moser, C.O.N. (1985). Housing Policy: Towards a Gender Awareness Approach. Working Paper No. 71. London: Development Planning Unit.

Papanek, H. (1982). "Purdah in Pakistan: Seclusion and Modern Occupations for Women", in H. Papanek and G. Minault (eds), *Separate Worlds: Studies of Purdah in South Asia*. Delhi: Chanakya Publications.

Singh, Atiya (1986). *A Child of Delhi*. UNFPA, 1986.

Timberlake, L., and J. Tinker (1984). *Environment and Conflict*. Earthscan Briefing Document No. 40. London: Earthscan.

Tinker, Irene, and Monique Cohen (1985). "Street Foods as a Source of Income for Women", in *Ekistics* 310, January/February 1985.

UN (1985). *Estimates and Projections of Urban, Rural and City Populations, 1950-2025: the 1982 Assessment.* New York: UN.

UNCHS (1986). Report of the Advisory Seminars on Women and Shelter. Vienna, 9-17 December 1985. Nairobi: UN Centre for Human Settlements.

UNEP (1985). *Sustainable Development and Peace.* Report prepared by the United Nations Environment Programme, Conference of the UN Decade for Women, Nairobi, 15-26 July 1985.

UNFPA (1986). *The State of the World Population, 1986.* Report of the UN Fund for Population Activities. New York.

WCED (1987). *Our Common Future.* Oxford University Press.

World Resources Institute and the International Institute for Environment and Development (1986). *World Resources 1986.* New York: Basic Books.

CHAPTER 7. WOMEN WORKING FOR CONSERVATION

IUCN (1980). *World Conservation Strategy: Living Resource Conservation for Sustainable Development.* IUCN, UNEP, WWF.

CHAPTER 8. TRAINING WOMEN

Allison, Helen, Georgina Ashworth and Nanneke Redclift (1986). *Hard Cash? Man-made development and its consequences: a feminist perspective on Aid.* London: Change, with support from War on Want, June 1986.

Ashworth, Georgina (ed) (1981). *Of Conjuring and Caring: Women in development.* London: Change.

Marker Kabraji, Aban (1987). *Pakistani Women's Study Group Tour to Indian Environmental Projects in the Garhwal-Kumaon Himalayas.* Report from IUCN-Pakistan.

Non-Governmental Liaison Service (1987). *Case Studies from Africa: Towards Food Security.* New York: UN.

Verghese, Valsa, Maria Teresa Chadwick and Ximena Charnes (1983). "Education and Communication: an overview", in *Women in Development. A Resource Guide for Organization and Action.* Geneva: ISIS.

Ward, Barbara and René Dubos (1972). *Only One Earth: The Care and Maintenance of a Small Planet.* Harmondsworth: Penguin Books.

World Bank, (1979). *Recognizing the "Invisible" Woman in Development: The World Bank's Experience.* Washington D.C.: The World Bank.

CHAPTER 9. PLANNING THE FAMILY

Agarwal, Anil (1985). "Population and Environment", in *Economic and Political Weekly*, Vol. 20 (24), June 15.

Hamand, J. (1987). "Fodder trees and family planning in Nepal", in "Earthwatch" section of *People*, Vol. 14 (3).

Huston, Perdita (1978). *Message from the Village*. New York: The Epoch Foundation.

IPPF (1985-1). *Annual Report*. London: IPPF.

IPPF (1985-2). *Experiences from Africa: Ghana, Kenya, Lesotho & Mauritius*. Nairobi: IPPF Africa Regional Office.

IPPF (1987). Information supplied by Frances Dennis, Director of Information.

IUCN (1984). *Population and Natural Resources*. Commission on Ecology Occasional Papers, no. 3. Gland, Switzerland: IUCN.

Makwavarara, Angelina (1984). "Women Realize Better than Anybody Else what an Accelerated Population Growth Rate Means ...", in *Ceres* (special issue on food, agriculture and women). Rome: FAO.

Salas, Rafael M. (1986-1). "Population and Sustainable Development", speech at the Conference on Conservation and Development: Implementing the World Conservation Strategy. Ottawa, Canada, 3 June 1986.

Salas, Rafael M. (1986-2). *UNFPA and International Population Assistance*. UNFPA, 1986. Reprint from *Harvard International Review*, March 1986.

Salas, Rafael M. (1987). *The State of World Population 1987*. UNFPA.

WCED (1987). *Our Common Future*. Oxford University Press.

CHAPTER 10. WOMEN ORGANIZE THEMSELVES

IWTC (1984). *Women Organizing: A collection of IWTC Newsletters on Women's Organizing and Networking Strategies*. New York: IWTC.

NGO Forum '85. Planning Committee (1986). Final Report of meetings in Nairobi, 10-19 July 1985. New York: IWTC, 1986.

CHAPTER 11. THE INTERNATIONAL RESPONSE

Commission on Ecology and Development Cooperation (1986). *Environment and Development Cooperation*. Report and Recommendations submitted to the Minister for Development Cooperation. The Hague: Government of the Netherlands.

Government of Denmark (1986). Paper on Danish Development Assistance.

ELC (1985). *Sustainable Development*. Report of the Proceedings of the Global

Meeting on Environment and Development for Non-Governmental Organizations, Nairobi, 4-8 February 1985. Nairobi: Environment Liaison Centre.

ELC (1986). *Women and the Environmental Crisis*. A Report of the Proceedings of the Workshop on Women, Environment and Development, Nairobi, 10-20 July 1985. Nairobi. Environment Liaison Centre.

IUCN (1980). *World Conservation Strategy: Living Resource Conservation for Sustainable Development*. IUCN/ UNEP/WWF.

IUCN (1985). *Implementing the World Conservation Strategy. IUCN's Conservation Programme 1985-1987*. IUCN Programme Series No. 1/1985.

IUCN (1986). *The IUCN Sahel Report: A Long-Term Strategy for Environmental Rehabilitation*. Report of IUCN's Task Force on the Sahel and other drought-affected regions of Africa. Gland, Switzerland: Conservation for Development Centre, IUCN.

Luangwa Integrated Resource Development Project (1987). *Proposals for the Phase 2 Programme*. LIRDP Project Document no. 3.

Mazza, Julia (1987). *The British Aid Programme and Development for Women*. London: War on Want Campaigns.

Ministerie Buitenlandse Zaken (1987). *Actieprogramma Vrouwen en Ontwikkeling*. The Hague, Netherlands: Directoraat-Generaal Internationale Samenwerking.

OECD (1983). *Guiding Principles to Aid Agencies for Supporting the Role of Women in Development*. Paris: OECD.

Tolba, Mostafa Kamal (1985). "An Alliance with Nature 'Women and the Earth's Traditions'", statement to World Conference to Review and Appraise the Achievements of the UN Decade for Women, Nairobi, 15-24 July 1985. Nairobi: UNEP.

CHAPTER 12. WORKING TOGETHER FOR THE FUTURE

Afsar, Haleh (1987). "Holding half the sky", in *A Women's World*. London: Thames Television.

Aidoo, Agnes Akosua (1985). *Women and Environmental Rehabilitation*. ATRCW/ECA.

Chambers, Robert (1987). *Sustainable Rural Livelihoods*. Overview paper for a Conference on Sustainable Development organized by the International Institute for Environment and Development, London.

Harrison, Paul (1987). *The Greening of Africa*. London: Paladin.

Mitter, Swasti (1986). *Common Fate, Common Bond*. London: Pluto Press.

Mukhopadhyay, Maitrayee (1984). *Silver shackles: women and development in India*. Oxford: Oxfam.

Rogers, B. (1983). "The Power to Feed Ourselves – women and land rights", in

Caldecott, L., and S. Leland (eds), *Reclaim the Earth*. London: The Women's Press.

Taylor, Debbie (1985). "Women: an analysis", in *Women: A World Report*. London: Methuen and New Internationalist.

Timberlake, Lloyd (1987). *Only One Earth – Living for the Future*. London: BBC/Earthscan.

WCED (1987). *Our Comon Future*. Oxford University Press.

Abbreviations

ADFG	=	Ação Democrática Feminina Gaúcha (Brazil)
AFOTEC	=	International Service for the Support of Training and Technologies in West Africa/Sahel (Senegal)
AWC	=	Associations of Women's Clubs (Zimbabwe)
CAP	=	Consumers' Association of Penang (Malaysia)
CDC	=	Conservation for Development Centre (IUCN)
CEMAT	=	Centro de Estudios Mesoaméricano sobre Tecnología Appropriada (Guatemala)
CHDSC	=	Centre for Human Development and Social Change (India)
CSE	=	Centre for Science and Environment (India)
DAC	=	Development Assistance Committee (OECD)
DANIDA	=	Danish International Development Agency
DAWN	=	Development Alternatives with Women for a New Era
DGSM	=	Dasohli Gram Swaraj Mandal (India)
ECA	=	Economic Commission for Africa (UN)
ELC	=	Environmental Liaison Centre
ENDA	=	Environment and Development in the Third World (regional)
EPOC	=	Equity Policy Center (USA)
FAFS	=	Fédération des Associations Féminines du Sénégal
FAO	=	Food and Agriculture Organization (UN)
FPA	=	Family Planning Association
IDWSSD	=	International Drinking Water Supply and Sanitation Decade
IFAD	=	International Fund for Agricultural Development (UN)
IFDP	=	Institute for Food and Development Policy (USA)
IFN	=	International Feminist Network
IIED	=	International Institute for Environment and Development
ILO	=	International Labour Organization
INADES	=	African Institute for Economic and Social Development
INSTRAW	=	International Research and Training Institute for the Advancement of Women
IPPF	=	International Planned Parenthood Federation
IUCN	=	International Union for Conservation of Nature and Natural Resources

IWTC	=	International Women's Tribune Center
JEEP	=	Joint Energy and Environment Projects (Uganda)
KENGO	=	Kenya Energy Non-Governmental Organization
KWAHO	=	Kenya Water for Health Organization
KWDP	=	Kenya Woodfuel Development Programme
MMD	=	Mahila Mangal Dal (India)
NCWK	=	National Council of Women of Kenya
NGOs	=	Non-Governmental Organizations
ODA	=	Overseas Development Administration (UK)
OECD	=	Organization for Economic Development
REECA	=	Regional Energy Conservation Association (East Africa)
SEWA	=	Self-Employed Women's Association (India)
SREP	=	Sudanese Renewable Energy Project
SSS	=	Sarvodaya Shramadana Sangamaya (Sri Lanka)
UNCHS	=	UN Centre for Human Settlements
UNDP	=	UN Development Programme
UNEP	=	United Nations Environment Programme
UNESCO	=	UN Educational, Scientific and Cultural Organization
UNICEF	=	UN International Children Emergency Fund
UNIFEM	=	UN Development Fund for Women
USAID	=	US Agency for International Development
WCED	=	World Commission for Environment and Development
WCS	=	World Conservation Strategy
WHO	=	World Health Organization
WICCE	=	Women's International Cross-Cultural Exchange
WORLDWIDE	=	World Women Working for Women Dedicated to the Environment
WRSM	=	Women's Revolutionary Socialist Movement (Guyana)
WWF	=	Worldwide Fund for Nature
YDD	=	Yayasan Dian Desa
YWCA	=	Young Women's Christian Association

RESOURCE ORGANIZATIONS ON WOMEN AND ENVIRONMENT

Ação Democrática Feminina Gaúcha (ADFG), FOE-Brazil,
C.P. 2617, Porto Alegre, RS 90.001, BRAZIL
(contact: María José Guazzelli)

Asian and Pacific Centre for Women and Development (APCWD),
c/o Asia and Pacific Development Centre, P.O. Box 2444,
Jalan Data, Kuala Lumpur, MALAYSIA

Chipko Information Centre,
Parvatiya Navjeevan Mandal,
P.O. Silyara, Pin 249 155, Tehri-Garhwal (U.P.), INDIA

Consumers' Association of Penang,
Ms. Evelyn Hong,
87 Cantonment Road, Penang, MALAYSIA

Development Alternatives with Women for a New Era (DAWN),
c/o Ms. Neuma Aguiar, Rua Paulino Fernandes, 32, Rio de Janeiro, RJ,
BRAZIL

Environment Liaison Centre (ELC),
Global Coalition for Environment and Development,
P.O. Box 72461, Nairobi, KENYA
(director: Shimwaayi Muntemba)

Equity Policy Center,
4818 Drummond Ave, Chevy Chase, MD 20815, USA

Green Belt Movement,
National Council of Women of Kenya,
prof. dr. Wangari Maathai,
P.O. Box 43741, Nairobi, KENYA

International Fund for Agricultural Development (IFAD),
1107, Via del Serafico, 00142 Rome, ITALY

International Labour Organization (ILO),
CH 1211 Geneva 22, SWITZERLAND

International Organization of Consumer Unions (IOCU),
Emmastraat 9, 2595 EG The Hague, NETHERLANDS

International Planned Parenthood Federation (IPPF),
Regent's College, Regent's Park, London NW1 4NS, ENGLAND

International Reference Centre for Water Supply and Sanitation (IRC),
P.O. Box 93190, 2509 AD The Hague, NETHERLANDS
(contact: Christine van Wijk-Sijbesma)

International Service for the Support of Training and Technologies
in West Africa/Sahel (AFOTEC),
Villag 8297, Sacre-Coeur 1, Dakar, SENEGAL

International Union for Conservation of Nature and Natural Resources
(IUCN),
Avenue du Mont Blanc, CH 1196 Gland, SWITZERLAND

IUCN Working Group on Women, Environment and Sustainable
Development,
Netherlands IUCN Committee,
Damrak 28-30, 1012 LJ Amsterdam, NETHERLANDS
(executive coordinator IUCN Working Group: Irene Dankelman)

International Women's Tribune Center (IWTC),
777 UN Plaza, New York, NY 10017, USA

ISIS/WICCE (Women's International Cross-Cultural Exchange),
P.O. Box 2471, 1211 Geneva 2, SWITZERLAND

Kenya Energy Non-Governmental Organizations (KENGO),
P.O. Box 48197, Nairobi, KENYA

Kenya Water for Health Organization (KWAHO),
P.O. Box 61470, Nairobi, KENYA

OXFAM,
274 Banbury Road, Oxford OX2 7DZ, ENGLAND

Research Foundation for Science, Technology and Natural Resources
Policy,
105 Rajpur Road, Dehra Dun, 248001, INDIA
(director: Vandana Shiva)

UNEP (UN Environment Programme),
Standing Committee of Senior Women on Sustainable Development,
P.O. Box 72, Nairobi, KENYA

UNIFEM (UN Development Fund for Women),
304 East 45th Street, Room 1106, New York, NY 10017, USA

Women's Resource Centre,
Shirkat-Gah, 1 Bath Island Road, Karachi-4, PAKISTAN

Women's Revolutionary Socialist Movement (WRSM),
44 Public Road, Kitty, Georgetown, GUYANA

World Neighbors,
5116 N. Portland Avenue, Oklahoma City, OK 73112, USA

WorldWIDE (World Women Working for Women Dedicated to the
Environment),
1718 P Street, N.W., suite 813, Washington, DC 20036, USA

YWCA (World Young Women's Christian Association),
37 Quai Wilson, 1201 Geneva, SWITZERLAND

CONTRIBUTORS

Dr Theresa C. Aloo, B Sc. M Sc., Lecturer in Forestry at the Egerton University College in Kenya. Conservator of Forests of the Kenya Forestry Department. Areas of expertise: forest protection and agroforestry. First woman to be appointed Assistant Conservator and Conservator of Forests in Kenya.

Dr Elizabeth Ardayfio-Schandorf, PhD., BA. Senior Lecturer, University of Ghana, Legon (Accra). Areas of expertise: rural development, women in development, rural energy systems, diffusion of innovation. Has been a consultant to the ECA, Addis Ababa, on women and rural energy systems.

Prahba Bhardwaj, freelance journalist. Member of the Association of Media Women in Kenya. Chairperson of "Waste Fighters", an environmental NGO in Kenya. Areas of expertise: women, environment and development, human resources, recycling and occupational health hazards.

Dr Ann Kramer Clark, Associate Professor of Philosophy. Visiting scholar at the Institute for Food and Development Policy, 1986-7. Areas of expertise: women's studies, politics of food, and philosophy. Her work is directed towards improving understanding between women worldwide.

Irene Dankelman, M Sc. in Biology. Project assistant of the Netherlands IUCN Committee. Coordinator of the project "both ENDS" – Environment and Development Service for NGOs – for Third World NGOs. Areas of expertise: Women, environment and development. Former coordinator, Netherlands IUCN Committee; she is currently coordinator of the IUCN Working Group on Women , Environment and Sustainable Development. Executed the project on "women, environment and sustainable development" which resulted in this book. Has published reports, papers and a Netherlands version of the World Conservation Strategy.

Joan Davidson, MA in Geography, MSc. in Conservation of Natural Resources. Research Fellow at the Barlett School of Architecture and Planning, University College London. Writer and researcher on environmental planning and conservation in the UK and the Third World, specializing in community-based environmental action. Former consultant to a consortium of

UK environmental organizations for the preparation of a UK National Conservation Strategy, and to IUCN for initial work on a National Conservation Strategy for Zambia, as well as to the UK Department of the Environment, OECD, UNESCO, IUCN and the European Foundation. Member: IUCN Commission on Sustainable Development. Author of three books and numerous papers, she is a regular contributor to the *Guardian* on the environment.

Nancy Fee has degrees in Development Studies, Population Growth and Social Administration. Currently, Programme Adviser on Women's Development for the International Planned Parenthood Federation (IPPF), London. Previous work includes women's development and family planning with Quaker groups and the Family Planning Association of Kenya, and the establishment of community health programmes in Uganda with the High Commission for Refugees.

María José Guazzelli, agricultural engineer. General coordinator of the Vacaria project, a demonstration farm and training centre in ecological agriculture in Brazil. Area of expertise: low external input agriculture. Co-author of the book *Agropecuaria sem Veneno* (Agriculture without Poison).

Aban Marker Kabraji, BSc in Biology. Director, WWF Regional Office, Pakistan. Responsible for managing joint IUCN/WWF projects, including educational programmes in Karachi, Pakistan's National Conservation Strategy and the women's study tour of India. Consultant for the IUCN/WWF marine turtle conservation programme, and on women's affairs for the Ford Foundation in Islamabad.

Sylvia Kreyenbroek, M Sc. in Geography. Research Assistant, Environmental Management Group (Twijnstra Gudde NV, management consultants, Deventer, Netherlands). Areas of expertise: physical geography, landscape ecology and land evaluation. Field experience in Sri Lanka and Kenya.

Meher K. Marker, BA (Hons) and MA in Sociology. Presently working on development projects with a specific focus on women. Project developer for the IUCN project "Protection of Mangroves through the formation of Cooperatives among Fishing Villages, along the Karachi coast", Pakistan. Areas of expertise: education, health, women and environment.

Joan Martin-Brown, BA in International Studies and Political Science. Senior Liaison Officer and Washington Representative, United Nations Environment Programme (UNEP). Founder and chairperson of WorldWIDE. Areas of expertise: environment and natural resource communications; public awareness strategies and public policy formulation; the development and implementation of activities in support of environmental education. Former president, the Bolton Institute for a Sustainable Future and director, the Office of Public Awareness of the US Environmental Protection Agency. Serves on the Board of Directors of Concern and is closely involved in the Global Tomorrow Coalition.

Dr Shimwaayi Muntemba, PhD Economic History with a focus on agricultural change and eco-development theories and systems. Director, Environment Liaison Centre. Areas of expertise: women in development, rural development, rural socio-economic change, with a focus on agriculture, environment, and policy analysis. Was coordinator of the programme "Food security, agriculture, forestry and environment" of the World Commission for Environment and Development. Former project manager, Rural Women's Programme of ILO, Geneva, and Senior Lecturer and Head of the Department of the University of Zambia. Has published several books.

Sybil Agatha Patterson, Senior Lecturer, Department of Sociology, Social Work Unit, University of Guyana. Coordinator of the Institute of Women's Studies of the University. Adviser on Senior Citizens' Welfare. Areas of expertise: education, appropriate technology, and community development. Working with the UN University on the Biogas Programme in Guyana.

Dr Carole Rakodi, BSc., PhD. Diploma in Town Planning. Lecturer, Department of Town Planning, University of Wales, Institute of Science and Technology (UK), where she coordinates courses on planning in developing countries. Member: Royal Town Planning Institute. Areas of expertise: Third World urban planning and housing policy, especially in the poorest areas. Former urban planner, Lusaka, Zambia (1971-8), World Bank consultant in charge of the Lusaka Housing Project Evaluation Team (1975-8), and consultant for the UK Overseas Development Administration in India.

Dr Vandana Shiva, Coordinator, Indian Research Foundation for Science, Technology and Natural Resource Policy; South Asian Seeds Action Network; and World Rainforest Movement. Areas of expertise: women and ecology, environmental policy, science, development and ecology. A people's scientist, working with ecology movements in rural India, especially the Chipko movement.

Dr Irene Tinker, Director and founder of the Equity Policy Center in the US and Professor of the International Development Programme, American University. Areas of expertise: differential impact on women of developments in agriculture, household energy, appropriate technology, food processing; and micro enterprises. Director, Office of International Science of the American Association for the Advancement of Science, and Assistant Director of ACTION. Founding member of women's research centres, women's organizations, and national women's professional groups.

Betha I. Berry Turner, Joint Director of AHAS Ltd, a UK consultancy bureau. Joint coordinator, Habitat International Council's NGO Habitat Project for IYSH '87. Has direct experience in the UK and US of organizing and running community groups, housing cooperatives and associations in inner-city areas with mixed ethnic groups. Research experience in self-managed housing in Mexico, Philippines, Tanzania, India and other Third World countries. Strong interest in community and individual learning through people's involvement in shared activities, such as planning and organizing their own houses and neighbourhood activities.

Renu Wadehra, Master in Development Studies, specializing in women and development. Research Assistant of the Netherlands Embassy's Women's Bureau in India. Areas of expertise: rural women and development, women and environment. Consultant to OXFAM (Indian Trust) on women's issues. Executed a case study on "women and afforestation" in a Himalayan village. Works at the grassroots level, especially with needy women.

Josette Marianne Wunder, teacher and Women's Liaison Officer of an environment-development project in Nepal. Member: Management Committee of AREA (Australian Association for Research and Environment Aid). Areas of expertise: appropriate technology, health and hygiene, sociology, conservation and development, grassroots action with village women and communities. She works towards equitable development based on environmental sustainability (especially as it relates to women), through research, practical involvement in rural development projects, lecturing and writing.

Ovril Amelia Yaw, teacher, national secretary, Women's Revolutionary Socialist Movement, Guyana. Areas of expertise: appropriate technology and management of economic projects. Co-ordinator, WRSM's Appropriate Technology programme.

INDEX

Abidjan, Ivory Coast,87
Ação Democrática Feminina Gaúcha
    (ADFG), Brazil, 25, 50 117, 120, 148–9
ADFG. See Ação Democrática Feminina
    Gaúcha
AFOTEC. See International Service for the
    Support of Training and Technologies in
    West Africa/Sahel
Africa, 4, 5, 9, 12, 16, 18, 21, 22, 30, 43, 45,
    68, 71, 74, 96, 100, 122, 128, 143, 163;
    East, 4, 77;
    North, 5, 9, 16;
    South, 16;
    West, 22–4, 45
African Institute for Economic and Social
    Development (INADES),
Afsar, Haleh, 172
Agarwal, Anil, 51, 91
Agarwal, Bina, 14, 69, 70, 71, 75
agriculture, 7–28, 121, 126, 127, 163
agroforestry, 56–7
Ahmed, Iftikhar, 19
Aidoo, Agnes, 174
Alcantara, Elsa, 69
Allison, Helen, 124
Aloo, Theresa, 43, 52–3, 65
Amazon basin, 11, 50
Amicos de Terra (Friends of the Earth,
    Brazil), 120
Anand, Anita, 51
Andes, 45, 68, 72
Angola, 32
Aranda, Josefina, 16
Ardayfio, Elizabeth, 68, 73, 74
Ardayfio-Schandorf, Elizabeth, 83
Arisan system, 95
Ashworth, Georgina, 123
Asia, 10, 12, 17, 31, 43, 45, 68, 74, 92, 100,
    128, 143;
    South, 100;
    South-East, 10, 100, 139
Association of Women's Clubs (AWC),
    Zimbabwe, 115
Australian Development Assistance Bureau,
    61, 165–6
Australian Association for Research and

Environmental Aid (AREA), 61, 63
AWC. See Association of Women's Clubs
Aziripe, Lourdes, 16

Bahrain, 161
Bahuguna, Sunderlal, 46
Bajracharya, Deepak, 79
Baldia Soakpit Pilot Project, 34, 95, 105–107,
    174
Bamba, Nonny, 13, 46
Bandyopadhyay, J., 10, 11
Bangladesh, 3, 8, 12, 70, 71, 78, 124, 139
Barnard, Geoffrey, 68, 70
Barry, Jessica, 21
Bart, P., 68
Behin, Mira, 118
Behin, Sarla, 118
Bharati, Sachidanand, 58, 60
Bhardwaj, Prabha, 39, 86
Bhatty, Zarina, 53
Bhopal, India, 89, 91, 98, 116, 163
Bhutan, 49
Biogas Programme, 84–5
Bolivia, 8, 51, 146
Botswana, 16, 126, 159
Braun, Arnelle, 21
Brazil, 8, 9, 17, 24–5, 34, 39–40, 45, 50, 117,
    120–21, 148–9
Britain, 147, 166–7, 168
Brundtland, Gro Harlem, 179
Brundtland Commission, 121
Buenos Aires, Argentina, 88
Bull, David, 11
Burkina Faso, 35–7, 52, 54, 67, 71, 73–4, 127,
    176
Burma, 45
Butalia, Subhadra, 92

Caceras, 79
Cairo, Egypt, 88
Cameroon, 13, 20
Canada, 147, 159
CAP. See Consumers Association of Penang
Cape Verde, 56, 104–105, 115
Caribbean, 5, 9, 16, 128, 143, 163
Castro, Giselda, 120

CDC. *See* Conservation for Development
    Centre
Celik, Aliye, 90
Cecelski, Elizabeth, 47, 48, 54, 56, 67, 68, 70,
    71, 72, 74, 76, 79, 90
CEMAT. *See* Centro de Estudios
    Mesoaméricano sobre Tecnologia
    Apropriada
Centre for Human Development and Social
    Change (CHDSA), India, 126
Central America, 15, 17, 79
Centro de Estudios Mesoaméricano sobre
    Tecnologia Apropriada (CEMAT),
    Guatemala, 79
Chad, 67
Chambers, Robert, 176
Chaney, Elsa, 20
Chauhan, Sumi Krishna, 40
Chavangi, Noel A., 55
CHDSC. *See* Centre for Human Development
    and Social Change
Chernobyl, 89
Chile, 12, 147
China, 10, 18, 31, 75, 85, 87
Chipko movement, 42, 48, 49, 50, 66, 115,
    118, 129–30, 174
Christian Aid, 168
CIDA, 33
Club of Madres, 78
Coalition Against Dangerous Exports, 149
Cohen, Monique, 95
Collins, Joseph, 10, 11
Colombia, 8, 9, 40–41
conservation, 113–14, 120, 156–7, 170, 172–3,
    177
Conservation for Development Centre (CDC),
    157, 158
Consumers' Association of Penang (CAP),
    Malaysia, 116, 122
Costa Rica, 8, 45
Courier, 67
Creevey, Lucy, 12
Cruz, Cerenalla A., 53
Cuanalo, M.E. de, 71, 78

DAC. *See* Development Assistance
    Committee
Dakar, Senegal, 67
Danish International Development Agency
    (DANIDA), 166
Dankelman, Irene, 17, 121
Dar es Salaam, Tanzania, 87
Dasohli Gram Swaraj Mandal (DESM), 50
Davidson, Joan, 96
DAWN. *See* Development Alternatives with
    Women for a New Era

dams, 35–7
DANIDA. *See* Danish International
    Development Agency
Davies, Miranda, 20
De Cuanalo, M.E., 71, 78
Deere, Carmen Diana, 20
deforestation, 22, 45–7, 53, 80, 90, 147, 161,
    163, 172
De Leal, Leon, 8
Delhi, India, 97–8
desertification, 31, 163, 172
Development Alternatives with Women for a
    New Era (DAWN), 143–4
Development Assistance Committee (DAC),
    169
Devi, Gurli, 59, 60
Devi, Shati Suresha, 118
disease and health problems, 32–3. 72, 88–91,
    98, 100, 120
Draper, Patricia, 43
Dominican Republic, 8, 20
drought, 30, 90, 115

Eckholm, Erik, 67, 70, 73, 74, 76
ecological processes, 114–15, 119
Ecuador, 8, 116, 145
education, 26, 96, 107, 113, 123–4, 129–30,
    134, 148
Egypt, 124, 139, 145
ELC. *See* Environment Liaison Centre
Elmendorf, M., 77
Elnur, Awatif Mahmoud, 77
El Salvador, 8, 45
ENDA. *See* Environment and Development in
    the Third World
energy, 4, 8, 66–86, 146, 147
Environment and Development in the Third
    World (ENDA), 116
Environment Liaison Centre (ELC), Kenya,
    117, 141, 149, 160–61, 176
environmental degradation, 89–90, 120, 172–3
EPOC. *See* Equity Policy Center (EPOC), 96
Eriksen, Mary Ann, 161
Esteros project, ??
Ethiopia, 27, 67, 72, 76, 77
Eviota, Elizabeth U., 16

FAFS. *See* Federation des Associations
    Feminines du Senegal
Family Health Advisory Services (FHAS),
    22–3
Family Planning Association (FPA), 134–5
FAO. *See* Food and Agriculture Organization
Far East, 5
farming, 12–15, 125

Farvar, Khadijek Catherine Razavi, 27
Federation des Associations Feminines du Senegal (FAFS), 115
Fee, Nancy, 139
fertilizers, 10, 15, 18, 47, 79, 84–5, 120
FHAS. *See* Family Health Advisory Services
Flanders, Stephanie, 37
Fleuret, Patrick C. and Anne, 69, 73
Foley, Gerald, 78
Foly, Ayele, 23
food, 4, 5, 7, 8–9, 19, 43, 78, 95–6, 108–9, 121, 125, 147, 163
Food and Agriculture Organization (FAO), 4, 7, 9, 28, 45, 47–8, 52, 53–7, 68, 87, 108, 131, 157, 168–9
forests and forestry, 42–65, 86, 115, 118–19, 121, 135
Fortmann, Louise, 56
Forum '85, 140–42, 175
Forward-Looking Strategies for the Advancement of Women (Nairobi Conference), 93–4, 154–6, 169, 175
Foster, Theodora, 12, 13, 52, 69, 73
Fowler, 11
FPA. *See* Family Planning Association
Fresco, Louise, 3, 9, 18
French, David, 56
Friends of the Earth, 25, 120, 149
fuel, 4, 8, 43, 66–86, 90
Fundación Natura, South America, 122

Gachukia, Dr Eddat, 141
Gambia, 21, 69
Gandhi, Indira, 49
Gandhi, Mahatma, 74
Gattegno, I., 74
Gebreselassie, Alasebu, 76
Germany, 85, 167
Ghana, 16, 18, 22, 23, 45, 68, 69, 73–4, 80–83, 91, 145, 146
Ghorepani project, 60–63
Golgotta Settlement Horticultural Initiative, 28
Greece, 147
Green Belt Movement, 51, 115, 141, 147–8, 151, 174
Green Revolution, 9–10, 14, 15, 24–5, 119, 120
GTZ (German Technical Agency), 85–6
Guatemala, 8, 67, 71, 79, 124
Guazzelli, Maria José, 10, 17, 25, 117, 149
Gubbels, Peter, 23
Guinea, 45, 54
Gupta, Vrinda, 74
Guyana, 79, 83–5, 149–50

Haiti, 45, 73

Halley, Ione, 83–5, 178
Hamand, Jorge., 88, 89, 92–3
Harrison, Paul, 176, 177–8
Hayes, P., 67
health, 26, 37, 39, 72, 88–9, 96, 99, 100, 107, 120, 121, 127, 132–5, 136–7, 146, 162, 163, 173
Himalayas, 45, 49, 57, 66, 69, 72, 115, 122
Hosken, Frau, 93, 94, 96
Hoskins, Marilyn W., 43, 52, 54, 55, 68, 71
housing, 92–5, 101–103
human settlements, 87–110
Human Settlements, Zambia, 109
Huston, 13, 134

Ibn-Badis, A., 123
ICDA,, 11
Iddi, Alice, 23
IDWSSD. *See* International Drinking Water Supply and Sanitation Decade
IFAD. *See* International Fund for Agricultural Development
IFN. *See* International Feminist Network
ILO. *See* International Labour Organization
INADES. *See* African Institute for Economic and Social Development
income-generating activities, 5–6, 39, 70, 73–74, 86, 92, 1378, 177
India, 3–4, 8, 10, 11, 14–15, 18, 30, 42, 45, 46–51, 52, 53, 54, 57–60, 68–9, 70, 72–3, 74–5, 88–9, 97–8, 115, 116, 118–19, 122, 126, 129–30, 139, 143, 145, 174
Indonesia, 8, 17, 45, 66, 77, 95, 136–9
Indonesian Planned Parenthood Association (IPPA), 137–9
INSTRAW. *See* International Research and Training for the Advancement of Women
International Coalition on Energy and Development, 149
International Drinking Water Supply and Sanitation Decade (IDWSSD), UN, 29, 34–5
International Federation of Organic Agriculture, 149
International Fund for Agricultural Development (IFAD), 21
International Feminist Network (IFN), 143
International Institute for Environment and Development (IIED), 92–3
International Labour Organization (ILO), 12, 56, 66, 69, 73, 76, 152, 178
International Organization of Consumer Unions; 149
International Planned Parenthood Federation (IPPF), 133, 134–5, 136–7
International Research and Training Institute

for the Advancement of Women
  (INSTRAW), 12, 29, 34, 125, 128–9, 141
International Service for the Support of
  Training and Technologies in West
  Africa/Sahel (AFOTEC), 126–8, 178
International Union for the Conservation of
  Nature and Natural Resources (IUCN),
  31, 37, 100, 113, 114, 117, 129, 130, 136,
  156–60, 170, 179
International Women's Tribune Center
  (IWTC), 140, 142
IPPA. See Indonesian Planned Parenthood
  Association
IPPF. See International Planned Parenthood
  Federation
Iran, 26–7, 30
Iraq, 9
irrigation, 10, 31
ISIS, 12, 17, 98, 142–3
Italy, 89
Ivory Coast, 13, 45, 46, 67, 126
IUCN. See International Union for the
  Conservation of Nature and Natural
  Resources
IWTC. See International Women's Tribune
  Center

Jakarta, Indonesia, 90
Jamaica, 95, 104–105
Jayal, Nalni, 31
JEEP. See Joint Energy and Environment
  Projects
Jiggins, Janice, 9
Joint Energy and Environment Projects
  (JEEP), 77

Kabondo Women's Group, 38–9
Kabraji, Aban Marker, 130
Kalahari, 43
KENGO. See Kenya Energy Non-
  Governmental Organization
Kenya, 13, 16, 32, 37–9, 51, 52–3, 63–5, 76–7,
  85–6, 115, 116, 122, 139, 145, 146, 147–8,
  167, 169, 172, 174, 178
Kenya Energy Non-Governmental
  Organization (KENGO), 51, 76–7, 122,
  141
Kenya Water for Health Organization
  (KWAHO), 37–9
Kenya Woodfuel Development Programme
  (KWDP), 55
Khamati, Beatrice, 77
Koenig, D., 73
Korea, 9, 52, 71, 139;
  South, 52
Kudat, A., 91

KWAHO. See Kenay Water for Health
  Organization
KWDP. See Kenya Woodfuel Development
  Programme

land, 7–8, 19. 70, 92, 96, 120, 172, 173
Laos, 45
Lappé, Frances Moore, 10, 11
Latin America, 5, 9, 11, 15, 21, 43, 45, 68, 104,
  128, 143, 145, 163
Lebanon, 147
Lesotho, 16, 52, 77
Leon de Leal, Magdalena, 8
Liberia, 45
Lingkod Tao-Kalikason (Secretariat for an
  Ecologically Sound Philippines), 50
Lorestan project, 26–7
Luangwa Integrated Resource Development
  Project, 170
Lusaka, Zambia, 92, 93, 96, 101–104, 108–10

Maathai, Professor Wangari, 51
Mabate Women's Groups, 95
McCall Skutch, Margaret, 48, 54
McLeod, Ruth, 104
Madagascar, 45, 77
Maendeleo Ya Wanawake, Women and
  Energy Project, 51, 85–6
Mahila Mangal Dal (MMD), India, 58–9
Makwavarara, Angelina, 132
Makokha, Juliet N., 86
Malawi, 4, 67, 73
Malaysia, 31, 45, 53, 89, 116, 159
Mali, 22, 23, 67, 73, 127
Mandals, 122
Manibog, F. 67
Marker, Meher K., 100
Marker Kabraji, Aban, 129
Martin-Brown, J., 144
Mazza, Julia, 167–8
Mexico, 8, 10, 16, 24, 38, 52, 71, 78, 90, 93, 95,
  133–34
Mexico City, 88, 89
Middle East, 9, 16, 21, 100, 163
Mitter, Swasti, 173
MMD. See Mahila Mangal Dal
Mombamba Women's Group, Kenya, 85–6
Morse, 14
Moser, C.O.N., 93
Moss, Patricia, 78
Mozambique, 13, 67
Mukhopadhyay, Maitrayee, 174
Muntemba, Shimwaayi, 117, 121-22
Musumba, Brazille, 51, 77
Muzira, Mary Tereza, 77

Nagbrahman, D., 69
Nairobi, Kenya, 87, 93, 94, 117, 140, 141, 175–6
Nash, June, 20
National Council of Women of Kenya (NCWK), 147
Nepal, 9, 45, 49, 52, 60–63, 66, 67, 68, 115, 135
Netherlands, 11, 41, 166
NGO Forum, 43, 94
Niamey, Niger, 67
Nicaragua, 9, 20, 139
Niger, 48, 67
Nigeria, 9, 45, 67
NORAD, 170
Nyoni, Sithembiso, 14, 18, 19

Obel, Elizabeth, 51
ODA. See Overseas Development Administration
OECD. See Organization for Economic Cooperation and Development
Organization for Economic Cooperation and Development (OECD), 169
Organization of Rural Associations for Progress (ORAP), Zimbabwe, 150–52
ORAP. See Organization of Rural Associations for Progress
Ouagadougou, Burkina Faso, 67
Overseas Development Administration (ODA), 166–7
Oxfam, 24, 28, 135, 168

Pakistan, 12, 30, 32, 34, 67, 76, 78, 90, 98–100, 105–107, 129–30, 139, 159, 174
Panama, 95
Papanek, H., 99
Papua New Guinea, 116, 147
Paraguay, 45
Paris, Thelma, R., 57
participation, women's, 28, 34, 52–3, 61–3, 102
Pastizzi-Ferencic, Dunja, 129
Pathah, Shekhar, 171
Patterson, Sybil, 85
Pearce, Jenny, 8
Perkumpulan Wanita Indonesia (PERWARI), 137–9
PERWARI. See Perkumpulan Wanita Indonesia
Peru, 5, 47, 69–70, 78
pesticides, 10, 15, 89, 120
Pesticides Action Network (PAN), 149
Philippines, 8, 16, 45, 50, 53, 95
Pinabetal Women's Organization, mexico, 24
Poldermans, Caroline, 15
pollution, 10, 31, 89, 115, 173

population growth, 80, 87, 90, 131–9, 147, 163, 175
Puerbo, Hasan, 66
purdah, 98–100

Quinlan, Susan, 14

Rakodi, Carole, 103, 110
Ramanankasina, Estelle, 46
REECA. See Regional Energy Conservation Association
reforestation, 47
refugees, 27–8, 76, 90
Regional Energy environment Conservation Association (REECA), East Africa, 77
Reiter, Rayna R., 43
Relief and Rehabilitation Commission (RRC), 27–8
Renner, Magda, 120
Revolutionary Ethiopian Women's Association (REWA), 28
Rocheleau, Dianne, 56
Rogers, B., 172
Ruether, Rosemary Radford, 176
Rwanda, 68, 71

Sachs, Karin, 16
Safa, Helen, 16, 20
Sahel, 30, 56, 57, 71, 90, 157
Sai, Dr Fred, 133
Salas, Rafael M., 131, 133
salinization, 10, 31
sanitation, 32, 33–5, 37, 39, 61–2, 79, 89, 91, 92–3, 101, 105–107, 135
Sao Paulo, Brazil, 88
Sarin, M., 77
Sarvodaya Shramadana Sangamaya (SSS), Sri Lanka, 166
Sathyamala, Dr, 98
Satterthwaite, David, 88, 89, 92, 102
Schuler, S., 68
Self-Employed Women's Association (SEWA), India, 21
Senegal, 67, 78, 95, 115, 116, 126–8
Seveso, Italy, 89
SEWA. See Self-Employed Women's Association
Shamba system, 63–4
Shanghai, China, 88
Sheth, Malini Chand, 46, 50
Shiva, Dr Vandana, 10, 11, 15, 19, 21, 42, 47, 49, 117–19
SIDA. See Swedish International Development Authority
Sierra Leone, 13, 16, 43, 146
Singh, Atiya, 97

Sinha, Rhadha, 8
SIRDO, 95
Skar, S.L., 70
Skutch, Margaret, 75
Slocum, Sally, 43
Smith, K.R., 72
social forestry, 47–8, 52, 53–5, 173
Somalia, 67, 68, 77
South Pacific, 147
Spears, J.S., 8, 48, 70
squatter settlements, 88, 101–102
SREP. See Sudanese Renewable Energy
   Project
Sri Lanka, 8, 11, 29, 71, 116
Srinavasan, V., 69
SSS. See Sarvodaya Shramadana Sangamaya
subsistence agriculture, 4, 9, 19
Sudan, 48, 57, 73, 77, 172
Sudanese Renewable Energy Project (SREP),
   77
sustainable agriculture, 19–28
sustainable development, 61, 81, 114, 121, 136,
   157, 174, 176
Sweden, 25
Swedish International Development Authority
   (SIDA), 25

Tanzania, 3, 18, 32, 48, 67, 68, 69, 77
Taylor, Debbie, 9, 15, 16, 17, 173
technology, 35, 60, 69, 106, 143, 147, 162, 163,
   164–5, 173
Thailand, 35, 45
Thomas, Lewis, 113
Thrupp, Lori-Ann, 52
Timberlake, Lloyd, 30, 176
Tinker, Irene, 69, 95, 96
Togo, 22–3
Tolba, Dr Mostafa K., 162–3
training, 38, 96, 120, 123–30, 178–9
Trinidad and Tobago, 146
Turner, Bertha, 105, 107

Uganda, 13, 77, 146
Ummaya, Pandurang, 49, 66
Union Carbide, 98
United Nations (UN), 4, 5, 21, 29, 30, 31, 84,
   85, 87, 93, 128, 148, 153, 154, 156, 177
United Nations Centre for Human
   Settlements (UNCHS), 91, 95, 160
United Nations Decade for Women, 38, 93–4,
   124, 140, 142, 153–69, 175–6, 179
United Nations Development Programme
   (UNDP), 34, 35, 37, 39, 56, 57, 136, 145,
   154
United Nations Environmental Programme
   (UNEP), 10, 12, 30, 89, 154, 160, 161–5,
   176

United Nations Educational, Scientific and
   Cultural Organization (UNESCO), 157
United Nations International Children
   Emergency Fund (UNICEF), 29 34, 38,
   39, 40, 107, 136, 141
UN Development Fund for Women
   (UNIFEM), 21, 165
United Nations Fund for Population
   Activities (UNFPA), 88, 97, 133, 136
United States of America (USA), 11, 145
United States Agency for International
   Development (USAID), 16, 52, 167
urban agriculture, 96, 108–10
urbanization, 87–90, 96

Vacaria Project, Brazil, 24–5, 149
Van Brakel, 18
Vancouver, Canada, 93
Van Wijk-Sybesma, 33, 34
Verghese, Valsa, 123, 124, 125

Wadehra, Renu, 60, 117
Walker, K.P., 159
Ward, Barbara, 125
water, 4, 8, 29–41, 72, 96, 110, 115–16, 127,
   163
WCED. See World Commission on
   Environment and Development
WCS. See World Conservation Strategy
Weiss, Ruth, 20
West German Aid Agency, 20
Whitehead, Ann, 15
WHO. See World Health Organization
WICCE. See Women's International Cross-
   Cultural Exchange
Wijkman, Anders, 30
Williams, Paula J., 47, 52, 54, 56, 78
Wisanti Utama, Elly, 77
Women in Forestry, 53, 64–5
Women's Construction Collective, 95,
   104–105
Women's International Cross-Cultural
   Exchange (WICE), 143
Women's Revolutionary Socialist Movement
   (WRSM), Guyana, 79, 141, 149–50
Women's World (ISIS), 98
Wood Energy Programme, 77
Working Women's Forum, 122
World Bank, 48, 76, 102, 123, 133, 146, 167
World Commission on Environment and
   Development (WCED), 86, 88, 96, 121,
   132, 171, 175, 179
World Conservation Strategy (WCS), 113–15,
   136, 158–60, 171, 175
World Consultation on Forestry Education,
   52

World Council of Churches, 141
World Health Organization (WHO), 10, 72, 134, 136, 154
World Neighbors, 22, 135
World Resources, 10, 30, 31, 43, 45, 55, 66, 67, 68, 87
WorldWIDE, 18, 140, 144–6, 179
Worldwide Fund for Nature. *See* World Wildlife Fund
World Wildlife Fund (WWF), 117, 157, 170, 173
WRSM. *See* Women's Revolutionary Socialist Movement
Wunder, Josette, 63
WWF. *See* World Wildlife Fund

Xavier Institute of Social Service, 126

Yaw, Ovril, 140, 150
Yayasan Dian Desa (YDD), 77
Y's Eye, 51, 146–7
Yemen, 73
Young Women's Christian Association (YWCA), 51, 146–7

Zaïre, 9
Zambia, 101–103, 108–10, 116, 147, 168, 170, 178
Zimbabwe, 13, 14, 18, 20, 95, 115, 150–52, 164, 178
Zonta Club, 38

For Product Safety Concerns and Information please contact our EU
representative GPSR@taylorandfrancis.com Taylor & Francis Verlag GmbH,
Kaufingerstraße 24, 80331 München, Germany

Printed and bound by CPI Group (UK) Ltd, Croydon, CR0 4YY

01/05/2025

01858580-0001